NEW ASPECTS OF BIOCHEMICAL PHYSICS: PURE AND APPLIED SCIENCE

SERGEI D. VARFOLOMEEV
ELENA B. BURLAKOVA
ANATOLII A. POPOV
AND
GENNADY E. ZAIKOV
EDITORS

Nova Science Publishers, Inc.
New York

For permission to use material from this book please contact us:
Telephone 631-231-7269; Fax 631-231-8175
Web Site: http://www.novapublishers.com

LIBRARY OF CONGRESS CATALOGING-IN-PUBLICATION DATA

New aspects of biochemical physics : pure and applied science / Sergei D. Varfolomeev ... [et al.] (editor).
p. ; cm.
Includes bibliographical references and index.
ISBN 1-60021-419-3
1. Physical biochemistry. I. Varfolomeev, Sergei Dmitrievich.
[DNLM: 1. Biochemical Phenomena. 2. Chemistry, Physical. QU 34 N532 2006]
QP517.P49N49 2006
572'.43--dc22 2006026104

Published by Nova Science Publishers, Inc. ✦ New York

CONTENTS

PREFACE

This book presents recent research in chemical and biochemical physics. Chemical physics addresses a large range of problems. An effective chemical physicist is a "jack-of-all-trades," able to apply the principles and techniques of the field to everything from high-tech materials to biology. Just as the fields of chemistry and physics have expanded, so have chemical physics subject areas, which include polymers, materials, surfaces/interfaces, and biological macromolecules, along with the traditional small molecule and condensed phase systems. Biochemical physics is a science that joins the three natural sciences (biology, chemistry and physics) into one comprehensive study. N.M. Emanuel pioneered this science over fifty years ago. This book brings together papers written by Emanuel's students, which reveal recent developments in this interesting field.

Chapter 1 - The enzyme PGHS carries out the first step in the biosynthesis of all prostaglandins, as well as thromboxane and prostacyclin, by catalyzing the conversion of an arachidonic acid molecule into prostaglandin H_2. PGHS carries out two reactions: the cyclooxygenase reaction, which is responsible for the conversion of the arachidonic acid into prostaglandin G_2, and the peroxidase reaction, which is responsible for the reduction of PGG_2 to the corresponding alcohol – prostaglandin H_2. Despite years of research, the kinetic mechanism of the action of PGHS is still not fully known: the absence of competition between and the cross activation of the two reactions cannot be explained on the basis of existing kinetic schemes. In the work reported here we have investigated the effect of the presence of an electron donor (TMPD) on the rate of the cyclooxygenase reaction. We suggest an alternative kinetic scheme for the action of the enzyme and a model for the competition between TMPD and arachidonic acid.

Chapter 2 - To determine the role of solvent polarity in the mechanism of action of biologically active compounds at ultra low concentrations we performed a comparative study of the effect of the natural antioxidant α-tocopherol (α-TP) prepared in a polar solvent (ethanol-aqueous solutions with predomination of water upon diluting) and in a nonpolar solvent (based on petrolatum oil), within a wide concentration range (10^{-4}M - 10^{-25}M) on the viscosity properties of the lipid bilayer of mouse hepatocyte microsomal membranes in vitro. The structural changes in membranes were detected on ESR-spectrometer "Bruker-200D" (Germany) by spin probe procedure with the use of stable nitroxyl radicals 5- and 16-doxylstearic acids to study the rigidity of the surface lipid bilayer regions (8 Å) and microviscosity of the deep membrane layers (20 Å), respectively. As a result nonlinear multimodal dose dependences of the effects of polar and nonpolar α-TP solutions, typical for the action at ultra low doses, have been obtained. It has been demonstrated that in the area of «apparent» α-TP concentrations ($<10^{-18}$ M) only polar (aqueous) α-TP solutions had a statistically reliable effect on the structure of microsomal membranes. This fact indicates on the key role of solvent polarity in the transfer of information about a compound in one or another form upon dilution in the course of preparation of solutions.

Chapter 3 – Using a homogeneous hydrophilic chemiluminescent model system the antiradical activity of 3-hydroxypyridine (3-HP) and 5-hydroxybenzimidazole (5-HB) derivatives was studied. By introducing of lipopolysaccharide endotoxin NO-inhibiting activity of 3-HP and 5-HB derivatives in mice liver was researched. It was estimated that some substances had antiradical activity and could change NO synthesis in mice liver. One of them, the 3-HP derivative (4-methylthyobenzimidazolile), had the highest antiradical and NO-inhibiting activity.

Chapter 4 - The method of research of phase diagram structures of reactive distillation process in systems with many chemical reactions is offered. Use of the modified variables offered by Zharov for a finding of special points on chemical reactive manifold and their types is entered.

Chapter 5 - There have been defined the optimal operating modes of the column fixed-bed catalyst reactor of dichloroacetic acid (DCAA) hydrogenation, providing both the high purity of commercial monochloroacetic acid (MCAA) and rational using of the expensive palladium activated charcoal-based catalyst. The most rational methods of sectioning and additional wetting of the reactor have been proposed.

Chapter 6 - There have been studied steady-state multiplicities in the continuous process of equilibrium open evaporation with an irreversible and reversible chemical reaction on the grounds of the software written in the *MatLab* environment. In case of an irreversible reaction there have been obtained results confirming deductions gained earlier under an analytical investigation of the similar process. It has been shown, that the type of a stable stationary state depends, besides operating parameters, on the system background. There has been gotten the criterion, permitting to predict an existence of several steady modes at the fixed process operating parameters on the base of information about system physicochemical properties.

Chapter 7 - There have been compared nine distillation-based sequences for the light gasoline C_{4-6+} fraction separation, including conventional and complex columns on basis of the criterion of the total heating steam and cooling water costs minimum. It has been shown that the direct distillation column arrangement is the optimal one for the considered feed composition and product fraction purities of 90.0–98.0 wt. %. It has been observed that the direct sequence remains optimal when varying the feed composition.

Chapter 8 - The motion of a magnetizable elastic body in an alternate magnetic field is experimentally studied. The body of a cylindrical from is located in a cylindrical channel. Traveling nonuniform magnetic field is generated by an electromagnetic system. Agreement of experimental data with theoretical results has been achieved.

Chapter 9 - The results of the interaction of highly dispersed magnetic particles with the surface of a foreign solid phase in different liquid media are given. On the basis of an analysis of experimental kinetic data, a phenomenological theory of the deposition process of magnetic particles on a foreign surface is developed. The dependences of the influence of the nature and composition of liquid media on the interaction of containing solid phases in the system under investigation are established. A stability criterion for magnetic particles during interaction with a foreign surface under different conditions is proposed.

Chapter 10 - In this study we used an electron spin resonance (EPR) method to explore the effect of thyrotropin-releasing hormone (TRH) in a wide range of concentrations (10^{-4} to 10^{-18} mol/l) on the structure of lipid bilayer of the plasmatic (PM) and endoplasmic reticulum (ER) membranes isolated from mice liver and brain cells. Using 3 stable free radicals localized in different regions of lipids as spin probes we have observed a great change in lipid microviscosity and order parameter under the effect of different TRH concentrations; the TRH concentration – effect dependence is nonlinear and polymodal. Comparing these changes with the results obtained in in vivo research carried out by I.P. Ashmarin and colleagues, we can conclude that: 1) the joint decreasing of order parameter in ER isolated from liver and brain cells is responsible for the TRH antiepileptic effect; 2) the sharp increasing (decreasing) of lipid microviscosity in PM isolated from liver (brain) under the effect of (10^{-9}-10^{-10}M) TRH is probably connected with interaction with TRH and its receptor; 3) modification of the of biological membrane structure is one of the mechanisms that TRH employs to achieve its anticonvulsant effect. In addition to that our results also support I.P. Ashmarin and colleagues' recommendation that clinically administered doses of TRH should be reduced.

Chapter 11 - For the first time it is executed quantum chemical calculation of a molecule 6-metilpergidrotetralin by method AB INITIO in basis 6-311 G ** with optimization of geometry on all parameters. The optimized geometrical and electronic structure of this connection is received. Acid force 6-metilpergidrotetralin is theoretically appreciated. It is established, that it concerns to a class of weak H-acids (pKa = + 35.85, pKa - a universal index of acidity), as pKa> 14.

Chapter 12 - For the first time it is executed quantum chemical calculation of a molecule 7-metilpergidrotetralin by method AB INITIO in basis 6-311 G ** with optimization of geometry on all parameters. The optimized geometrical and electronic structure of this connection is received. Acid force 7-metilpergidrotetralin is theoretically appreciated. It is established, that it concerns to a class of weak H-acids (pKa = + 35.31, pKa - a universal index of acidity), as pKa> 14.

Chapter 13 - For the first time it is executed quantum chemical calculation of a molecule of gopan, being one of the major intermediate products in the most complicated evolutionary process of natural synthesis of oil, method AB INITIO in basis 6-311 G ** with optimization of geometry on all parameters. The optimized geometrical and electronic structure of this connection is received. Acid force gopan is theoretically appreciated. It is established, that it concerns to a class of very weak H-acids (pKa = + 33, a pKa-universal index of acidity), since pKa> 14.

Chapter 14 - For the first time it is executed quantum chemical calculation of a molecule of diagopan, being one of the major intermediate products in the most complicated evolutionary process of natural synthesis of oil, method AB INITIO in basis 6-311 G ** with optimization of geometry on all parameters. The optimized geometrical and electronic structure of this connection is received. Acid force diagopan is theoretically appreciated. It is established, that it concerns to a class of very weak H-acids (pKa = + 34.7, a pKa-universal index of acidity), since pKa> 14.

Chapter 15 - Based on the established mechanism of catalysis by nickel complexes $Ni(II)(L^1)_2$ (L^1=acac$^-$), activated by additives of electron-donor monodentate ligands-modifiers L^2 (L^2=HMFA, DMFA, MP, MSt) in the process of selective alkylarens (ethylbenzene and cumene) oxidation with dioxygen into corresponding hydroperoxides (PEH) the effective methods for control of the selective ethylbenzene oxidation into α-phenylethylhydroperoxide (PEH), including the use of quaternary ammonium salts R_4NBr as L^2 exoligands-modifiers are suggested by us. Values of selectivity, S_{PEH}, conversion degree, C, and PEH yield reached at application of L^2=R_4NBr (Me_4NBr) exceed analogous parameters in the presence of the other {Ni(II)$(L^1)_2$+L^2} systems and known catalysts of ethylbenzene oxidation into PEH. The principle possibility of the catalysts construction on the basis of complexes Fe(III)$(acac)_3$ and R_4NBr for the increase in rate and selectivity of the ethylbenzene oxidation by dioxygen into PEH is demonstrated. The proposed "dioxygenase-like" mechanism of the transformation of iron catalyst into more active catalytic particles in the course of oxidation process is discussed. The value method of activity of nickel and iron catalysts in the micro steps of the chain-radical ethylbenzene oxidation is offered.

Chapter 16 – It was shown that the scale or direction of correlations between both the several parameters of the lipid composition and the generalized parameters of the phospholipids composition of the murine liver depend on the kinetic characteristics of their lipids (the lipid peroxides content, the capacity of lipids to decompose peroxides). It was found that the combined intraperitoneal administration of the low- toxicity surfactant (Tween 80) and acetone at low doses causes disturbances in the lipid peroxidation regulatory system which manifest themselves in 1 month after the action.

Chapter 17 - An analysis of published data on the radiolytic properties of chitosan and some modelling its fragments substances was performed. The main process of radiation chitosan degradation is connected with the formation and conversion of free radicals. The mechanisms of the primary radicals $\dot{C}2$, $\dot{C}1$ and $\dot{C}3$ (of the NH_2 and H removal radicals) conversion up to the formation of some end products of chitosan radiolysis were offered in the form of the generalized schemes. Problem of the DNA and membrane protection in irradiated cell is discussed.

In: New Aspects of Biochemical Physics
Editors: Sergei D. Varfolomeev et al., pp. 1-16
ISBN 1-60021-419-3

Chapter 1

KINETIC MECHANISM OF THE ACTION OF THE BIFUNCTIONAL ENZYME PROSTAGLANDIN-H-SYNTHASE. THE INHIBITING ACTION OF TMPD AS AN ELECTRON DONOR ON THE CYCLOOXYGENASE REACTION

***L. A. Tsaplina*[1], *Y. O. Karatasso*[2],**
***I. S. Filimonov*[2] *and P. V. Vrzheshch*[1]**

[1]International Research Centre for Biochemical Technology
M.V. Lomonosov Moscow State University
1 building 12 Mendeleevskaya Str., Moscow 119992, Russia
[2]N.M. Emanuel Institute of Biochemical Physics,
Russian Academy of Science, 4 Kosygin Str., Moscow 119991, Russia

ABSTRACT

The enzyme PGHS carries out the first step in the biosynthesis of all prostaglandins, as well as thromboxane and prostacyclin, by catalyzing the conversion of an arachidonic acid molecule into prostaglandin H_2. PGHS carries out two reactions: the cyclooxygenase reaction, which is responsible for the conversion of the arachidonic acid into prostaglandin G_2, and the peroxidase reaction, which is responsible for the reduction of PGG_2 to the corresponding alcohol – prostaglandin H_2. Despite years of research, the kinetic mechanism of the action of PGHS is still not fully known: the absence of competition between and the cross activation of the two reactions cannot be explained on the basis of existing kinetic schemes. In the work reported here we have investigated the effect of the presence of an electron donor (TMPD) on the rate of the cyclooxygenase reaction. We suggest an alternative kinetic scheme for the action of the enzyme and a model for the competition between TMPD and arachidonic acid.

Key Words: PGHS, enzyme kinetics, competitive inhibition

The enzyme PGHS catalyzes the conversion of arachidonic acid into PGH_2, which is a universal precursor of prostaglandins and thromboxanes. The synthesis of PGH_2 from arachidonic acid is carried out in two steps [1,2]:

1. The conversion of arachidonic acid into an organic peroxide – prostaglandin G_2 – as a result of a cyclooxygenase reaction.
2. The reduction of prostaglandin G_2 into prostaglandin H_2 as a result of a peroxidase reaction.

Prostaglandin H_2 can then be converted under the action of specific convertases into physiologically active compounds such as prostaglandin E_2, D_2, $F_{2\alpha}$, thromboxane A_2, prostacyclin [3-7]. These compounds play an important role in the growth of tumor cells [8], in blood clotting [9-10], and in the development of bronchial asthma [11]. These compounds also function as neuromodulators [12,13].

Therefore the enzyme PGHS is a key link in the synthesis pathway of intracellular regulators, which supply all the subsequent stages with universal raw material – prostaglandin H_2. Mammals possess two isoforms of PGHS: PGHS-1 (which is always present in cells and which ensures their normal functionality) [43] and PGHS-2 (which is produced in response to particular stimuli such as cytokines and growth factors) [44]. In the present work we have studied the properties of PGHS-1 taken from vesicular glands of sheep.

In 1977, a method for isolating of a solubilized PGHS from vesicular glands was developed. Active research of the kinetic characteristics of prostaglandin synthesis ensued. Using hemin as a cosubstrate [14,15], PGHS catalyzes two reactions: cyclooxygenase and peroxidase. Since the catalysis of both reactions takes place on the same enzyme molecule, the question of how these two activities are related arises. It has been shown that the exhibition of enzyme's cyclooxygenase activity requires the presence of trace amounts of peroxide in the reaction mixture. [16,17] The removal of peroxides from the solution via the addition of glutathione-peroxidase and the reducted glutathione leads to the appearance of a time lag on the kinetic curve of the absorption of oxygen during the cyclooxygenase reaction. The lag disappears if one adds peroxides exogenously into the solution [49,50]. This means that during the course of the peroxidase reaction a certain active form of the enzyme is produced, which can carry out the cyclooxygenase catalysis.

At the same time it was shown that free radicals play a role in cyclooxygenase reaction [18]. In another work [19] it was shown that during the course of the peroxidase reaction a radical of the tyrosine amino acid residue is produced; and it is that radical that initiates the cyclooxygenase reaction. By now, this hypothesis has been substantially proved [46-48]. To explain these observations, in 1988, the first mechanism, called as "branched-chain" mechanism for the action of the enzyme has been suggested (Figure 1). According to this mechanism, after the reduction of a peroxide molecule, the enzyme gets either two electrons from an external donor or one electron from an amino acid residue of tyrosine (producing a tyrosine radical Tyr*, as a result) and the other electron – from an external donor.

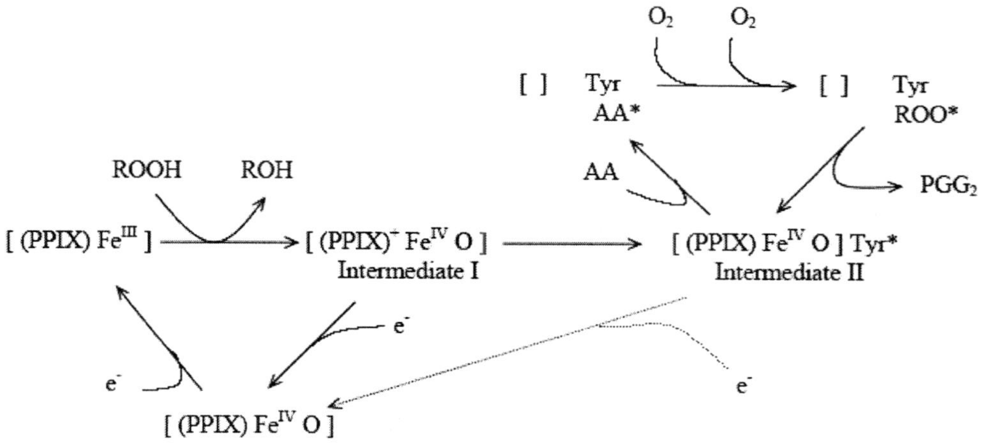

Figure 1. Branched-chain mechanism for the action of PGHS.

Despite the fact that this scheme explains the activation of cyclooxygenase catalysis by a peroxidase reaction, it has a serious drawback. The scheme implies that an enzyme molecule can participate either in the cyclooxygenase reaction, in which case it cannot catalyze the peroxidase reaction, or in the peroxidase reaction, in which case it cannot catalyze the cyclooxygenase reaction. In other words, these two processes compete with each other, so that an increase in the concentration of the substrate for one of the two reactions should lead to a decrease in the rate of the other. However, this effect has not been observed in practice. In fact several experiments such as the work [21] have shown that, on the contrary, the rate of the cyclooxygenase reaction increases while the concentration of electron donors increase. To explain this result, which does not fit within the framework of the "branched-chain" mechanism [39-41], alternative mechanisms have been suggested: "tightly-coupled" mechanism [39-41], shown in Figure 2, and "combined" mechanism [26], shown in Figure 3.

In the framework of these mechanisms, one can explain the increase of the rate of the cyclooxygenase reaction with the increased concentration of electron donor. However, these mechanisms have their drawbacks, as well. In the work [22], a qualitative and quantitative comparison of the predictions of the "branched-chain" and the "tightly-coupled" mechanisms was carried out. It was shown that several features can be explained only in the framework of the "branching-chain" mechanism. In particular, the "tightly-coupled" mechanism cannot explain the observed [22] accumulation of prostaglandin G_2 during the course of the reaction as well as the stoichiometry between the consumption of electron donor and arachidonic acid. (The measured ratio of the concentration of the utilized electron donor to the concentration of the oxidized form of the arachidonic acid was equal to 1.3 while the "tightly-coupled" mechanism predicts a value of 2.0). Moreover, it is clear that the "tightly-coupled" and "combined" mechanisms predict the absence of cyclooxygenase reaction if no electron donors are present in the solution, which contradicts experimental results in the literature [23,24,32-34] as well as the experimental results of this work. The "combined" mechanism does predict an increase in the rate of the cyclooxygenase reaction with the concentration of the electron donor increase; it cannot, however, explain the accumulation of prostaglandin G_2 [22] during the course of the reaction. Therefore, each of the three suggested mechanisms has significant drawbacks, and neither of them can strictly describe the kinetics of the PGHS reaction.

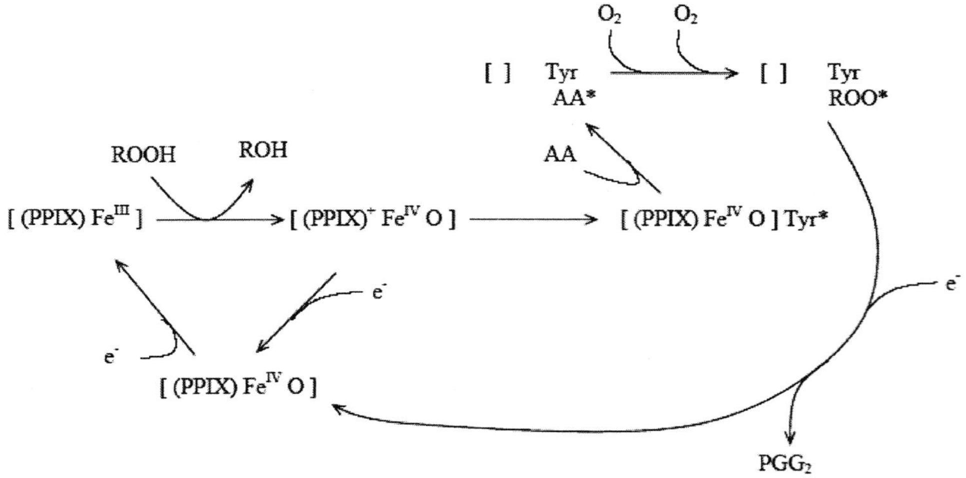

Figure 2. Tightly-coupled mechanism for the action of PGHS.

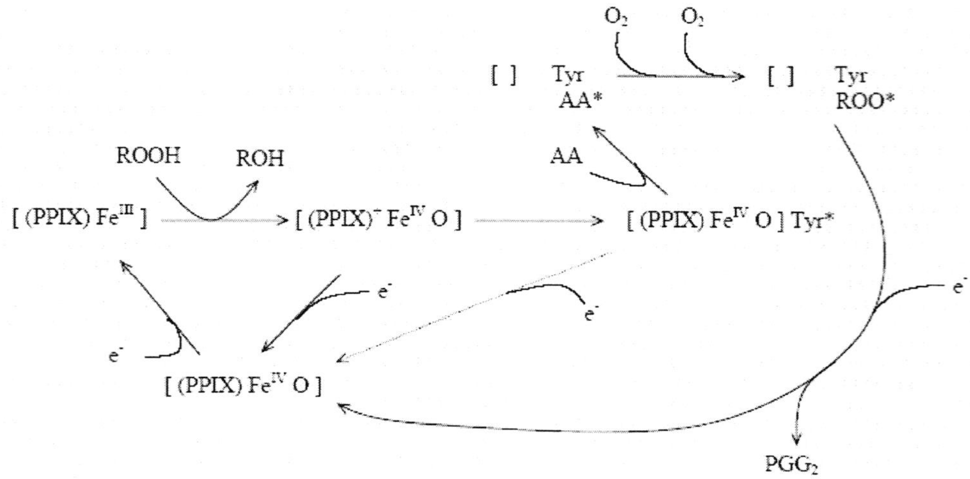

Figure 3. Combined mechanism for the action of PGHS

At the same time the mathematical analysis of the kinetic models for the action of the PGHS enzyme has been attempted [37,38]. However, due to the complexity of the system under investigation, the authors had to resort to significant simplifications, which led to the loss of important features of the PGHS kinetics.

Separately the mechanisms of cyclooxygenase and peroxidase reactions have been studied in sufficient detail. However, the issue of their interplay remains open since not only cyclooxygenase produce a substrate for the peroxidase but also the two reactions take place on the same molecule, which may be the origin of several key features of the PGHS reaction kinetics.

The aim of this work was the study of the kinetics interplay between the peroxidase and the cyclooxygenase reactions of PGHS using the polarographic and spectrophotometric methods of investigation. The word “interplay” here refers to the effect of the concentration of the substrate of one of the reactions on the rate of the other. In the present work, we examined the effect of an electron donor (TMPD) – a substrate for the peroxidase reaction – on the initial rate of the cyclooxygenase and the overall reactions (to be defined below). TMPD was chosen as the electron donor since it is a well-known one-electron donor [45]. The absorption maximum of its oxidized form is at 611 nm, (the extinction coefficient of the oxidized form of TMPD at this wavelength is 13500 $M^{-1}\cdot cm^{-1}$ [29]) which is far from the absorption region of hemin (that is the cofactor of PGHS) near 400 nm.

MATERIALS AND METHODS

Tris(hydroxymethyl)methylamine, tween-20, TMPD were bought from MP Biomedicals Inc.; arachidonic acid and hydrogen peroxide were bought from Sigma-Aldrich. All the other high purity reagents were made in Russia.

PGHS was isolated from vesicular glands of sheep using the method described in reference [25].

All the experiments used the following buffer: 50 mM Tris-HCl, 0.1% tween (i.e. 1 ml of twin-20 were added per 1000 ml of buffer), which is the standard buffer used in the study of PGHS kinetics [18-20]. The temperature of the reaction mixture was always 25°C. Reagents were added to the buffer and the reaction was then initiated by the addition of the enzyme. The arachidonic acid and TMPD were stored and added to the reaction mixture in the form of a solution in ethanol.

The flow of the peroxidase reaction and the flow of the overall reaction were monitored spectrophotometrically by observing the accumulation of the oxidized form of TMPD. This method allows to obtain the full kinetic curve in real time. The volume of the reaction mixture was 2.5 ml while the volume of the added enzyme solution was 10 ul. (The concentration of the protein was 4.5 mg/ml.) In order to monitor the overall PGHS reaction, arachidonic acid, TMPD, and the enzyme were added to the buffer.

$$AA + 2O_2 + 2TMPD + 2H^+ \rightarrow PGH_2 + 2TMPD^+$$

In order to monitor the overall PGHS reaction in the presence of hydrogen peroxide, hydrogen peroxide was added to the buffer in addition to arachidonic acid, TMPD, and the enzyme.

$$AA + 2O_2 + 2TMPD + 2H^+ \rightarrow PGH_2 + 2TMPD^+$$
$$H_2O_2 + 2TMPD + 2H^+ \rightarrow 2H_2O + 2TMPD^+$$

In order to monitor the flow of the peroxidase reaction, hydrogen peroxide, TMPD, and the enzyme were added to the buffer.

$$H_2O_2 + 2TMPD + 2H^+ \rightarrow 2H_2O + 2TMPD^+$$

The flow of the cyclooxygenase reaction was monitored amperometrically using the Clark electrode. This method of observation also allows to obtain the full kinetic curve in real time. The volume of the reaction mixture was 1.7 ml while the volume of the added enzyme solution was 10 ul. (The concentration of the protein was 4.5 mg/ml.) In order to monitor the cyclooxygenase reaction, TMPD, arachidonic acid, and the enzyme were added to the reaction mixture.

$$AA + 2O_2 \rightarrow PGG_2$$

The mathematical analysis of the data was carried out using the data analysis software Origin 6.1.

EXPERIMENTAL RESULTS

The stimulation (at low concentrations) and the inhibition (at high concentrations) of the cyclooxygenase reaction by various electron donors were observed in several other works [23,24,32-34]. These works, however, were mostly phenomenological, and did not make use of kinetic models. (This effect does not take place in the presence of any electron donor: electron donors such as L-adrenalin, ABTS, ferrocyanide do not produce this effect in a wide range of concentrations, as shown in Figure 5). In order to construct and quantitatively analyze a model for the inhibition of the cyclooxygenase reaction by an electron donor, we carried out a series of kinetic experiments.

Our experiments showed that an increase of the concentration of TMPD to a particular magnitude at various concentrations of the arachidonic acid causes an increase in the rate of the cyclooxygenase reaction, while further increase of the TMPD concentration decreases the rate of cyclooxygenase reaction (see Figure 4).

The stimulation of cyclooxygenase reaction at low concentrations and its inhibition at high concentration has been described in the literature for many donors such as MK-447 [32], phenol [23], guaiacol [33], acetaminophen [34], dimethylphenylenediamine [33], methylphenylsulphide [24] and others. In addition, several hypotheses regarding the possible causes of such inhibiting effects of the donors have been suggested. Firstly, these effects can be related to the fact that the presence of donors in the solution leads to the non-enzymatic depletion of the peroxide [33], which is an activator of cyclooxygenase. Secondly, donors may interact with radicals, such as the tirosyl-radical, whose presence in the active site of the cyclooxygenase is necessary for the cyclooxygenase reaction to proceed. Thirdly, donors may competitively interact with the substrate-binding site of cyclooxygenase [35]. It is also possible that the observed effect is caused by a combination of some or all of these effects.

It has been shown, (data not presented) that the addition of hydrogen peroxide at saturating concentration (140 mM) does not eliminate the inhibiting effect of TMPD. This means that the hypothesis that suggests the decrease of peroxide, which is an activator for the cyclooxygenase, at high concentrations of the electron donor is not applicable in this case.

From the hypothesis that TMPD interacts (non-competitively with respect to AA) with the active site of the cyclooxygenase by lowering the concentration of the tyrosyl-radical, it follows that a change in the concentration of the arachidonic acid should not have any effect on the inhibiting action of TMPD. (It means, that maximum velocity is achieved at the constant concentration of TMPD independently of the concentration of arachidonic acid)

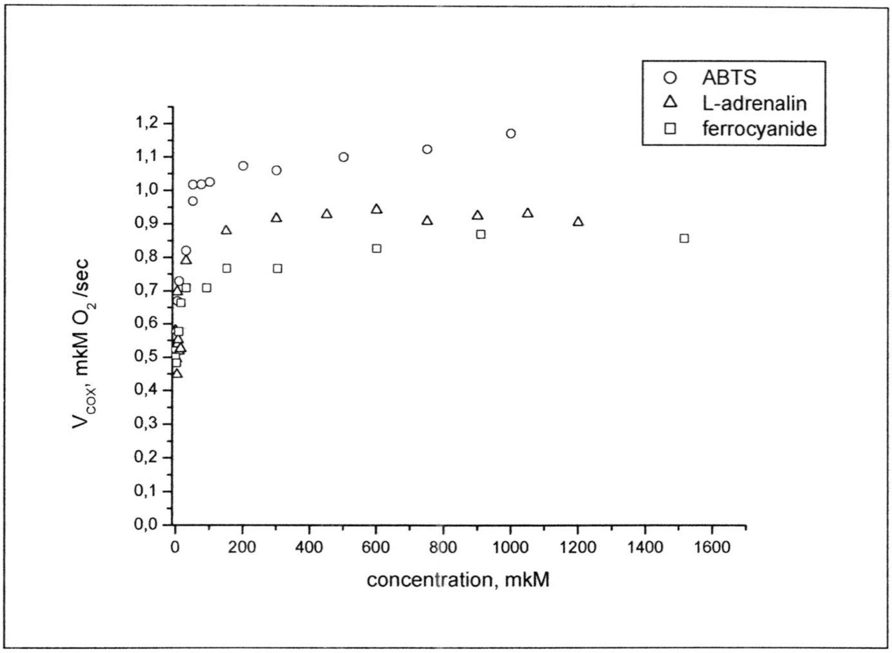

Figure. 4. The dependence of the initial rate of the cyclooxygenase reaction on the concentration of TMPD at various (40 uM, 75 uM, 100 uM, 180 uM) concentrations of the arachidonic acid. Conditions are standard (see Materials and methods). The solid curves are the result of approximation of the experiments according to the Eq. [20].

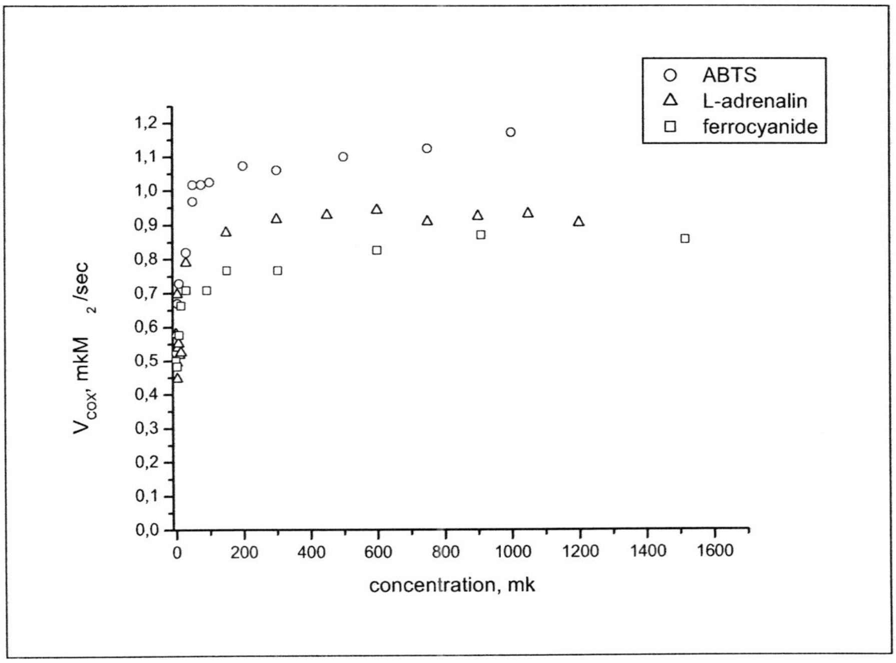

Figure. 5. The dependence of the initial rate of the cyclooxygenase reaction on the concentration of L-adrenalin, ABTS, ferrocyanide. Concentration of arachidonic acid is 100 mkM. Conditions are standard and the hemin at concentration of 2 mkM was added.

However, in practice an increase in the concentration of the arachidonic acid leads to a maximal reaction rate at higher concentrations of TMPD (see Figure 4). Therefore, this hypothesis should be discarded.

Therefore, the only remaining hypothesis for the explanation of this effect is the one based on a competition between AA and TMPD during their interaction with the substrate-binding site of cyclooxygenase.

We have also studied independently the dependence of the initial rate of the cyclooxygenase reaction at a fixed concentration of TMPD on the concentration of the arachidonic acid (see Figure 6). This figure shows that to first order the dependence is linear in the double-reciprocal coordinates. Based on this we can conclude that the interaction of the enzyme with the arachidonic acid can be described in the framework of the Michaelis-Menten model.

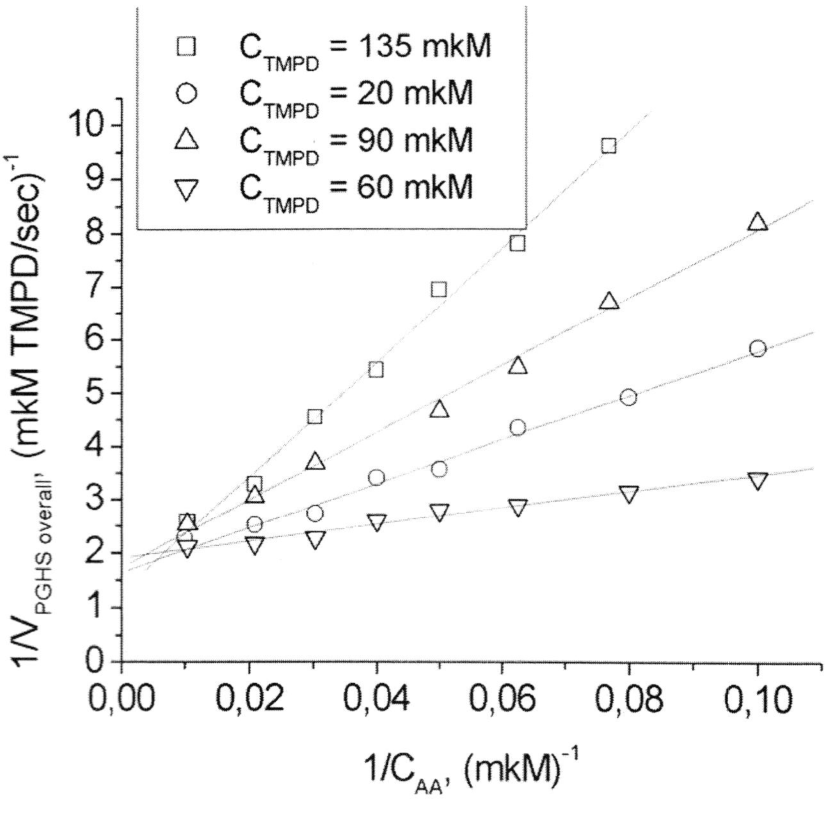

Figure. 6. The dependence of the initial velocity of the cyclooxygenase reaction at fixed concentrations of TMPD on the concentration of the arachidonic acid. Conditions are standard.

The Construction of Kinetic Models and Discussion

As shown above, the existing "branched-chain", "combined", and "tightly-coupled" models cannot describe all the experimental data available for this enzyme. Therefore, it is necessary to suggest an alternative mode for the action of PGHS, a model that can explain all the observed kinetic features of the enzyme.

To construct the new model, we consider the general model for the kinetic action of an enzyme that catalyzes two reactions and uses several substrates in the process. Applying with the general scheme [36,42,51] to the description of the action of PGHS, we constructed a scheme with the minimal number of nodes that reflects the known features of the kinetics of this enzyme. This scheme is shown in Figure 7.

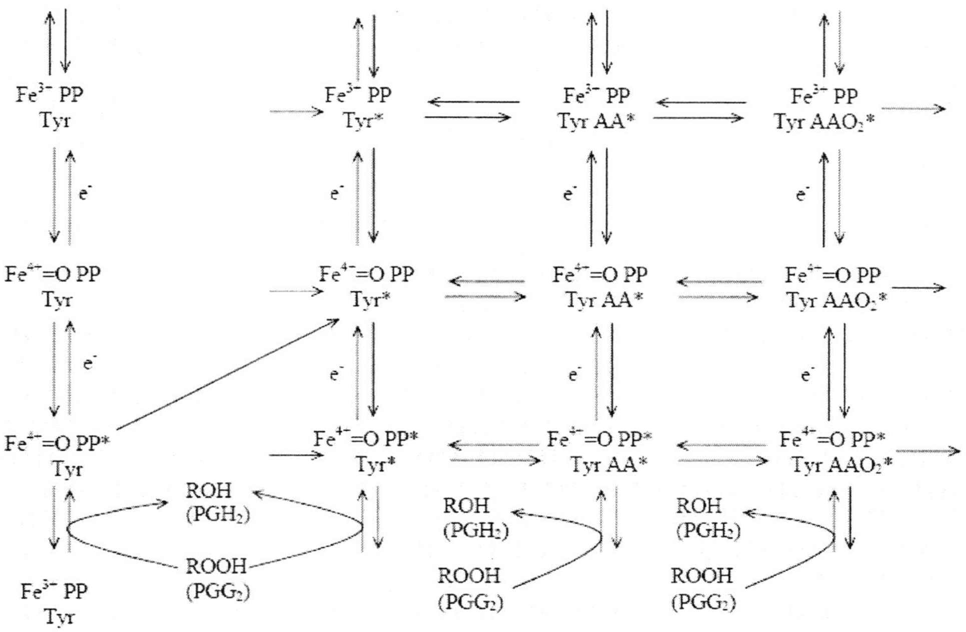

Figure. 7. The suggested scheme for the action of PGHS.

Since the cyclooxygenase reaction requires the presence of tyrosyl-radical (Tyr*), which can be produced during the peroxidase reaction, we have to allow in the scheme for the presence of the enzyme without Tyr*, which cannot carry out the cyclooxygenase catalysis. During the course of the peroxidase catalysis, the enzyme loses two electrons (from the protoporphyrin ring and from the iron atom) to the reduction of the peroxide. In order to reduce the next peroxide molecule, the enzyme has to get back the two electrons. It either take both electrons from an external donor (such as, for example, TMPD), or it can take one electron from the aminoacid residue of tyrosin and the other – from an external donor. The second scenario is the one that leads to the production of Tyr*.

It is very hard to carry a rigorous mathematical analysis of this scheme. However, the production of tyrosyl-radical, which is necessary for the cyclooxygenase reaction, takes place very quickly (it appears only 1-2 seconds after the reaction begins) and does not affect the steady-state rate of the reaction [30,31]. This allows for significant simplification of the calculations, which, according to Ref. [36] lead to the following dependence of the initial steady-state rate of the cyclooxygenase reaction on the concentration of the electron donor – the substrate of the peroxidase reaction:

$$v^{COX} = \frac{e_0 + e_1 \cdot [D] + e_2 \cdot [D]^2 + e_3 \cdot [D]^3 + e_4 \cdot [D]^4}{d_0 + d_1 \cdot [D] + d_2 \cdot [D]^2 + d_3 \cdot [D]^3 + d_4 \cdot [D]^4} \quad (1)$$

While constructing kinetic models for the description of our data, we also neglected the presence of the enzyme without a tyrosyl-radical since the time lag on the kinetic curves was not observed in our experiments (data not presented).

For certain ratios between constants e_i and d_i, this formula describes the stimulation of the cyclooxygenase reaction by the presence of higher electron donor concentration. At the same time, the model in Figure 7 does not contradict the accumulation of prostaglandin G_2 in the reaction mixture. Therefore, our new model describes the known features of PGHS kinetics better than the "branched-chain", "tightly-coupled", and "combined" mechanisms. However, as already noted above, the ability to inhibit the cyclooxygenase reaction depends on the nature of the electron donor and is not common to all the donors. Consequently, it is necessary to build a separate model for the description of this effect.

To this end, a simplified model has been suggested, in which it is assumed that the rate of the cyclooxygenase reaction catalyzed by the enzyme depends on the state of the active site of the peroxidase reaction. It is also assumed that the cyclooxygenase reaction does not affect the rate of the peroxidase reaction. This is confirmed, for example, by the fact that the presence in the reaction mixture of non-steroid anti-inflammatory drugs that inhibit the cyclooxygenase effect of the enzyme does not affect the flow of the peroxidase reaction [52-54]. The kinetic scheme of the two reactions is shown in Figures 8a and 8b.

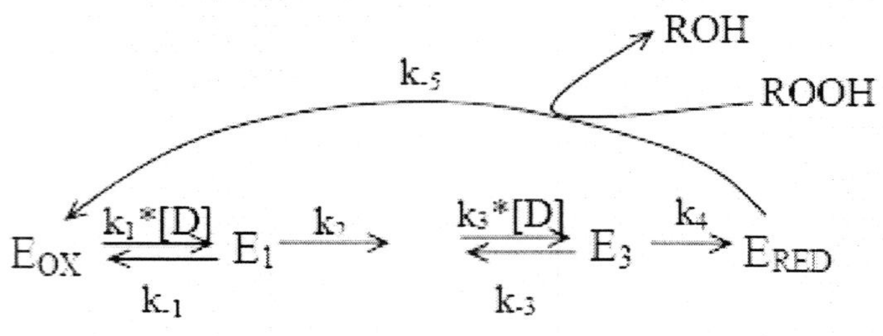

Figure 8a. Scheme for the peroxidase reaction

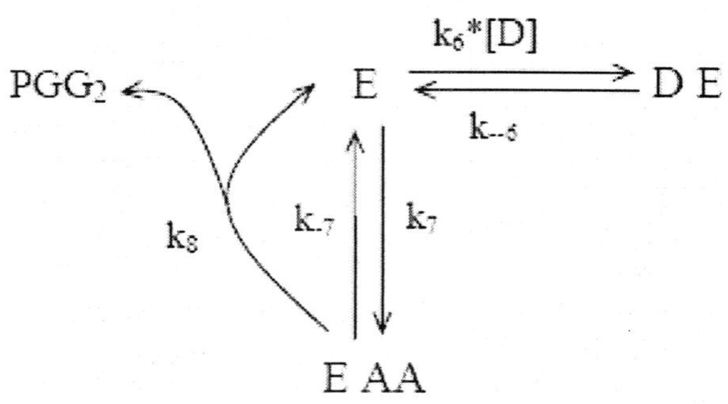

Figure 8b. Scheme for the cyclooxygenase reaction

Consider the peroxidase reaction. Here D denotes the electron donor TMPD; E_{red} and E_{ox} denote the reduced and oxidized forms of the enzyme; E_1 and E_3 denote the enzyme-substrate complex with the electron donor while E_2 denotes the half-reduced form of the the enzyme. It is assumed that each of the intermediate forms of the enzyme can catalyze cyclooxygenase reaction. Let us now compute the steady-state concentration of each form of the enzyme:

$$[E_{OX}]=[E_0]\cdot\frac{(k_{-1}+k_2)\cdot k_3\cdot k_4\cdot k_5}{K_1+K_2\cdot[D]} \quad (2)$$

$$[E_1]=[E_0]\cdot\frac{k_1\cdot k_3\cdot k_4\cdot k_5\cdot[D]}{K_1+K_2\cdot[D]} \quad (3)$$

$$[E_2]=[E_0]\cdot\frac{(k_{-3}+k_4)\cdot k_1\cdot k_2\cdot k_5}{K_1+K_2\cdot[D]} \quad (4)$$

$$[E_3]=[E_0]\cdot\frac{k_1\cdot k_2\cdot k_3\cdot k_5\cdot[D]}{K_1+K_2\cdot[D]} \quad (5)$$

$$[E_{RED}]=[E_0]\cdot\frac{k_1\cdot k_2\cdot k_3\cdot k_4\cdot[D]}{K_1+K_2\cdot[D]} \quad (6)$$

where E_0 is the total enzyme concentration, [D] is the electron donor concentration, while constants K_1 and K_2 can be expressed in terms of rate coefficients as follows:

$$K_1=(k_{-1}+k_2)k_3k_4k_5+(k_{-3}+k_4)k_1k_2k_5 \quad (7)$$

$$K_2=k_1k_3k_4k_5+k_1k_2k_3k_5+k_1k_2k_3k_4 \quad (8)$$

Equations (2-6) imply that the intermediate forms of the enzyme can be subdivided into two groups based on the dependence of their concentration on the concentration of electron donors. E_{ox} and E_2 for one group, while E_{red}, E_1, and E_3 form the other. Defining

$$[E^1]=[E_1]+[E_3]+[E_{RED}] \quad (9)$$

$$[E^2]=[E_{OX}]+[E_2] \quad (10)$$

we find

$$[E^1] = \frac{K_2 \cdot [D]}{K_1 + K_2 \cdot [D]} \tag{11}$$

$$[E^2] = \frac{K_1}{K_1 + K_2 \cdot [D]} \tag{12}$$

Suppose now that the rate coefficients of the cyclooxygenase catalysis are equal for the enzyme forms of the same group. We can then denote them with superscripts 1 and 2 for the first group and the second group, respectively.

Let us now consider the cyclooxygenase reaction and take into account the fact that the electron donor can compete with the arachidonic acid while binding in the substrate-binding-site of the cyclooxygenase. Let us make the following definitions: v_{cox} – the observed steady-state rate of the cyclooxygenase reaction; v_1 – the steady state rate of the cyclooxygenase reaction, catalyzed by the forms of the enzyme from the first group; v_2 - the steady state rate of the cyclooxygenase reaction, catalyzed by the forms of the enzyme from the second group. Then

$$v^{COX} = v^1 + v^2 \tag{13}$$

Computing the dependence of v_1 and v_2 on rate coefficients and concentrations of substrates, we obtain

$$v^1 = [E^1] \cdot \frac{k_8^1 \cdot [AA]}{[AA] + \frac{k_{-7}^1 + k_8^1}{k_7^1} \cdot \left(1 + \frac{k_6^1}{k_{-6}^1} \cdot [D]\right)} \tag{14}$$

$$v^2 = E^2 \cdot \frac{k_8^2 \cdot [AA]}{[AA] + \frac{k_{-7}^2 + k_8^2}{k_7^2} \cdot \left(1 + \frac{k_6^2}{k_{-6}^2} \cdot [D]\right)} \tag{15}$$

Suppose now that we are in the quasistatic limit, i.e. for each form of the enzyme $k_8 << k_{-7}$. Then equations (12) and (13) can be written in the form

$$v^1 = E^1 \cdot \frac{k_8^1 \cdot [AA]}{[AA] + \frac{k_{-7}^1}{k_7^1} \cdot \left(1 + \frac{k_6^1}{k_{-6}^1} \cdot [D]\right)} \tag{16}$$

$$v^2 = [E^2] \cdot \frac{k_8^2 \cdot [AA]}{[AA] + \frac{k_{-7}^2}{k_7^2} \cdot \left(1 + \frac{k_6^2}{k_{-6}^2} \cdot [D]\right)} \tag{17}$$

Suppose also that for each form of the enzyme all the rate coefficients for the cyclooxygenase reaction are equal except for k_8, which determines the rate of accumulation of prostaglandin G_2. Then

$$v^1 = [E^1] \cdot \frac{k_8^1 \cdot [AA]}{[AA] + \frac{k_{-7}}{k_7} \cdot \left(1 + \frac{k_6}{k_{-6}} \cdot [D]\right)} \tag{18}$$

$$v^2 = [E^2] \cdot \frac{k_8^2 \cdot [AA]}{[AA] + \frac{k_{-7}}{k_7} \cdot \left(1 + \frac{k_6}{k_{-6}} \cdot [D]\right)} \tag{19}$$

Finally, using Eq. (13), we find

$$v^{COX} = \frac{K_1 \cdot V_2 + K_2 \cdot V_1 \cdot [D]}{K_1 + K_2 \cdot [D]} \cdot \frac{[AA]}{[AA] + K_5 \cdot (1 + K_6 \cdot [D])} \tag{20}$$

According to Eq. (20), the dependence of the initial rate of the cyclooxygenase reaction on the concentration of arachidonic acid is linear in double-reciprocal coordinates, which agrees with experimental data (Figure 6). Eq. (20) contains six constants. The software package Origin 6.1 was used to adjust the values of these constants to make Eq. (20) the best fit to the experimental data. The results of this approximation are given in Table 1.

Table 1. Mean values and standard deviation of constants V_1, V_2, K_1, K_2, K_5, K_6, found based on the analysis of experimental data

Constant name	Constant dimension	Mean value	Deviation
V_1	mkM/sec	8.8	1.5
V_2	mkM/sec	0.48	0.1
K_1	$mkM^{-2} \cdot sec^{-4}$	26	5
K_2	$mkM^{-3} \cdot sec^{-4}$	0.29	0.05
K_5	mkM	1.8	0.5
K_6	mkM^{-1}	1.1	0.2

The main conclusion that follows from the data presented in this table is that depending on the state of the active site of the peroxidase reaction, the rate of the cyclooxygenase reaction can change by a factor of 10 to 15.

In conclusion, in this work we have suggested an alternative scheme for the action of PGHS, a scheme that describes all the known kinetic features of this enzyme more adequately that the classic schemes. Moreover, we have studies the inhibiting effect of an electron donor such as TMPD, which is a substrate of the peroxidase reaction, on the initial rate of the cyclooxygenase reaction. It was shown that the only hypothesis that can describe the observed effects is the one based on the competition between TMPD and the arachidonic acid for the substrate-binding site of the cyclooxygenase reaction. Finally, we have analyzed a simplified model that takes into account the competition between TMPD and the arachidonic acid as well as the main idea of the suggested alternative model, which consists of the fact that both reactions catalyzed by PGHS can take place on a single enzyme molecule simultaneously.

REFERENCES

[1] M. Hamberg, B. Samuelsson, *Proceedings of National Academy of Science of the United States of America*, Vol. 70, N. 3, pp. 899-903, 1973.

[2] D.H. Nugteren, E. Hazelhof, *Biochimica et Biophysica Acta*, Vol. 326, N 3, pp. 448-61, 1973.

[3] E. Christ-Hazelhof, D.H. Nugteren, *Biochimica et Biophysica Acta*, Vol. 572, N. 1, pp. 43-51, 1979.

[4] S. Bergstrom, H. Danielsson, B. Samuelsson, *Biochimica et Biophysica Acta*, Vol. 90, pp. 207-10, 1964.

[5] K. Watanabe, R. Yoshida, T. Shimizu, O. Hayaishi, *Advances in Prostaglandin Thromboxane and Leukotriene Research*, Vol. 15, pp. 151-5, 1985.

[6] P. Wlodawer, S. Hammarstrom, *FEBS Letters*, Vol. 97, N. 1, pp. 32-6, 1979.

[7] V. Ullrich, M. Haurand, *Advances in Prostaglandin Thromboxane and Leukotrine Research*, Vol. 11, pp. 105-10, 1983.

[8] G.S. Eisenbarth, D.K. Wellman, H.E. Leibovitz, *Biochemical and Biophysical Research Communications,* Vol. 60, N. 4, pp. 1302-08, 1974.

[9] C. Malmsten, M. Hamberg, J. Svensson, B. Samuelson, *Proceedings of National Academy of Science of the United States of America*, Vol. 72, N. 4, pp. 1446-50, 1975.

[10] G.P. McNicol, *Lancet*, Vol. 2, N. 8197, pp. 736-8. 1980.

[11] G.J. Dusting, S. Moncada, J.R. Vane, *Advances in Prostaglandin Thromboxane and Leukotrine Research,* Vol. 10, pp. 59-106, 1982.

[12] J.F. Burka, R.J. Flower, *British Journal of Pharmacology*, Vol 65, N. 1, pp. 35-41, 1979.

[13] T. Shimizu, M. Muzino, T. Amano, O. Hayaishi, Proceedings of National Academy of Science of the United States of America, Vol. 76, N. 12, pp. 6231-4, 1979.

[14] D.P. Hemker, J.W. Aiken, Prostaglandins, Vol. 20, N. 2, pp. 321-32, 1980.

[15] G.J. Roth, E.T. Machuga, P. Strittmatter, *Journal of Biological Chemistry*, Vol. 256, N. 19, pp. 10018-22, 1981.

[16] M.E. Hemler, C.G. Crawford, W.E. Lands, *Biochemistry*, Vol. 17, N. 9, pp. 1772-9, 1978.

[17] M.E. Hemler, W.E. Lands, *Journal of Biological Chemistry*, Vol. 255, N. 13, pp. 6253-61, 1980.

[18] R.P. Mason, B. Kalyanaraman, B.E. Tainer, T.E. Eling, *Journal of Biological Chemistry*, Vol. 255, N. 11, pp. 5019-22, 1980.

[19] R. Karthein, R. Dietz, W. Nastainczyk, H.H. Ruf, *European Journal of Biochemistry*, Vol. 171, N. 1, pp. 313-20, 1988.

[20] R. Dietz, W. Nastainczyk, H.H. Ruf, *European Journal of Biochemistry*, Vol. 171, N. 2, pp. 321-8, 1988.

[21] Y. Hsuanyu, H.B. Dunford, *Journal of Biological Chemistry*, Vol. 267, N. 25, pp. 17649-57, 1992.

[22] R.J. Kulmacz, A.L. Tsai, C. Wei, *Biochemistry*, Vol. 34, N. 26, pp. 8499-512, 1995.

[23] R.W. Egan, J. Paxton, F.A. Kuehl, *Journal of Biological Chemistry*, Vol. 251, N. 23, pp. 7329-35, 1976.

[24] R.W. Egan, P.H. Gale, E.M. Baptista, K.L. Kenicott, W.J. VandenHeuvel, R.W. Walker, P.E. Fagerness, F.A. Kuehl, *Journal of Biological Chemistry*, Vol. 256, N. 14, pp. 7352-61, 1981.

[25] F.J. Van der Ouderaa, M. Buytenhek, D.H. Nugteren, D.A. Van Dorp, *Biochimica et Biophysica Acta,* Vol. 487, N. 2, pp. 315-331, 1977.

[26] V. Koshkin, B.H. Dunford, *Biochimica et Biophysica Acta*, Vol. 1430, N. 2, pp. 341-8, 1999.

[27] R.J. Kulmacz, A.L. Tsai, G. Wu, *Biochemistry*, Vol. 42, N. 46, pp. 13772-7, 2003.

[28] R.J. Kulmacz, B. Bambai, *Journal of Biological Chemistry*, Vol. 275, N. 36, 27608-14, 2000.

[29] R.J. Kulmacz, Y. Ren, A.L. Tsai, G. Palmer, *Biochemistry*, Vol. 29, N. 37, pp. 8760-71, 1990.

[30] R.J. Kulmacz, P.J. Marshall, W.E. Lands, *Journal of Biological Chemistry*, Vol. 262, N. 8, pp. 3510-7, 1987.

[31] Tsai, G. Wu, G. Palmer, B. Bambai, *Journal of Biological Chemistry*, Vol. 274, N. 31, pp. 21695-700, 1990.

[32] F.F. Kuehl, J.L. Humes, R.W. Egan, E.A. Ham, G.C. Beveridge, C.G. Van Arman, *Nature*, Vol. 265, N 5590, pp. 170-3, 1977.

[33] R.W. Egan, P.H. Gale, G.C. Beveridge, L.J. Marnett, F.A. Kuehl, *Advances in Prostaglandin Thromboxane and Leukotrine Research,* Vol. 6, pp. 153-5, 1980.

[34] J. Robak, A. Wieckowski, R. Gryglewski, *Biochemical Pharmacology*, Vol. 27, N. 4, pp. 393-6, 1978.

[35] J.L. Humes, C.A. Winter, S.J. Sadowski, F.A. Kuehl, *Proceedings of National Academy of Science of the United States of America,* Vol. 78, N. 4, pp. 2053-6, 1981.

[36] P.V. Vrzheshch, Moscow, *"Stationary kinetics of multi-substrate enzyme reactions"*, MAKS Press, 2003, 56 pp.

[37] J.H. Tien, W.D. Hazelton, R. Sparks, C.M. Ulrich, *Bulletin of Mathematical Biology*, Vol. 67, N. 4, pp. 683-700, 2005.

[38] W.D. Hazelton, J.H. Tien, V.W. Donato, R. Sparks, C.M. Ulrich, *Biochemical Pharmacology*, Vol. 68, N. 3, pp. 423-32, 2004.

[39] Y. Hsuanyu, H.B. Dunford, *Journal of Biological Chemistry*, Vol. 267, N. 25, pp. 17649-57, 1992.

[40] M. Bakovic, M, H.B. Dunford, *Biochemistry*, Vol. 33, N. 37, pp. 6475-82, 1994.

[41] M. Bakovic, H.B. Dunford, *Journal of Biological Chemistry*, Vol. 271, N. 4, pp. 2048-56, 1996.

[42] P.V. Vrzheshch, E.A. Batanova, A.T. Mevkh, S.D. Varfolomeev, I.G. Gazaryan, R.N. Thorneley, *Biochemical Journal*, Vol. 372, N. 3, pp. 713-24, 2003.

[43] C. Yokoyama, T. Tanabe, T. Biochemical and Biophysical Research Community, Vol. 165, N. 2, pp. 888-94, 1989

[44] D.A. Jones, D.P Carlton, T.M. McIntyre, G.A. Zimmerman, S.M. Prescott, *Journal of Biological Chemistry*, Vol. 268, N. 12, pp. 9049-54, 1993.

[45] R.J. Kulmacz, *Prostaglandins*, Vol. 34, N. 2, pp. 225-40, 1987.

[46] D.C. Goodwin, M.R. Gunther, L.C. His, B.C. Crews, T.E. Eling, R.P. Mason, L.J. Marnett, *Journal of Biological Chemistry*, Vol. 273, N. 15, pp. 8903-9, 1998.

[47] T. Shimokawa, R.J. Kulmacz, D.L. DeWitt, W.L. Smith, *Journal of Biological Chemistry*, Vol. 265, N. 33, pp. 20073-6, 1990.

[48] Tsai, L.C. His, R.J. Kulmacz, G. Palmer, W.L. Smith, *Journal of Biological Chemistry*, Vol. 269, N. 7, pp. 5085-91, 1994.

[49] M.E. Hemler, G Graff, W.E. Lands, *Biochemical and Biophysical Research Communications*, Vol. 85, N. 4, pp. 1325-31, 1978.

[50] M.E. Hemler, W.E. Lands, *Journal of Biological Chemistry*, Vol. 255, N. 13, pp. 6253-61, 1980.

[51] P.V. Vrzheshch, *Biochemistry (*Mosc), Vol. 64, N. 4, pp. 502-512, 1999.

[52] T. Miyamoto, N. Ogino, S. Yamamoto, O. and Hayaishi, *Journal of Biological Chemistry,* Vol. 251, N. 9. pp. 2629-36, 1976.

[53] G.J. Roth, C.J. Siok, *Journal of Biological Chemistry*, Vol. 253, N. 11. pp. 3782-84, 1978.

[54] J.A. Lindgren, H.E. Claesson, S. Hammarstrom, S. *Prostaglandins*, Vol. 13, N. 6, pp. 1093-1102, 1977.

In: New Aspects of Biochemical Physics
Editors: Sergei D. Varfolomeev et al., pp. 17-22
ISBN 1-60021-419-3

Chapter 2

The Role of Solvent Polarity in the Mechanism of Action of Biologically Active Compounds at Ultra Low Concentrations

V. V. Belov, E. L. Mal'tseva, N. P. Pal'mina and E. B. Burlakova
N.M. Emanuel Institute of Biochemical Physics, Russian Academy of Sciences
4, Kosygin str., Moscow 119991, Russia

Abstract

To determine the role of solvent polarity in the mechanism of action of biologically active compounds at ultra low concentrations we performed a comparative study of the effect of the natural antioxidant α-tocopherol (α-TP) prepared in a polar solvent (ethanol-aqueous solutions with predomination of water upon diluting) and in a nonpolar solvent (based on petrolatum oil), within a wide concentration range (10^{-4}M - 10^{-25}M) on the viscosity properties of the lipid bilayer of mouse hepatocyte microsomal membranes in vitro. The structural changes in membranes were detected on ESR-spectrometer "Bruker-200D" (Germany) by spin probe procedure with the use of stable nitroxyl radicals 5- and 16-doxylstearic acids to study the rigidity of the surface lipid bilayer regions (8 Å) and microviscosity of the deep membrane layers (20 Å), respectively. As a result nonlinear multimodal dose dependences of the effects of polar and nonpolar α-TP solutions, typical for the action at ultra low doses, have been obtained. It has been demonstrated that in the area of «apparent» α-TP concentrations ($<10^{-18}$ M) only polar (aqueous) α-TP solutions had a statistically reliable effect on the structure of microsomal membranes. This fact indicates on the key role of solvent polarity in the transfer of information about a compound in one or another form upon dilution in the course of preparation of solutions.

Keywords: α-tocopherol, solvent polarity, ultra low doses, spin label technique, microviscosity, structural characteristics of the membrane lipids.

Introduction

Ample experimental data accumulated to date provide evidence for the effect of various classes of biologically active compounds (BAC) at ultra low concentrations at different organizational levels of biological systems—from macromolecules, organs, and tissues to animals, plants, and even populations [1, 2]. We regard those BAC concentrations ultra low, the efficiency of which cannot be explained in terms of common standpoints and requires developing new methods. In our considerations, we proceeded from the maximal affinity of ligands for a receptor and took concentrations of 10^{-13} M and smaller as ultra low.

Each compound may have its own target and metabolic characteristics; however, there are similarities in the effect of all compounds at ultra low concentrations [2]:

(1) Nonmonotonous, nonlinear multimodal concentration-effect dependence. In most cases, the maximal activity is observed at certain concentration ranges, separated from one another by the so-called "dead area", where the effect is not expressed.
(2) Changes in the sensitivity (usually an increase) of biological objects to various agents (both endo- and exogenous).
(3) Manifestation of kinetic paradoxes, i.e., the possibility of detecting the effect of BAC at ultra low concentrations when the same compound is present in the cell or organism at the concentrations that exceed the ultra low concentration by several orders of magnitude. The effect on a receptor of a compound at the concentrations that are by several orders of magnitude lower than the dissociation constants of the ligand-receptor complex.
(4) The dependence of the "sign" of an effect on the initial characteristics of object.
(5) "Segregation" of properties of BACs upon a decrease in its concentration, when its activity is still retained, but side effects disappear.

The nature of these consistent patterns of the effect of BACs at ultra low concentration may be accounted for by the commonness of critical targets (e.g., cellular and subcellular membranes). A number of studies, some of which were performed in our laboratory, showed that the introduction of BACs at ultra low concentrations dissolved in solvents of different polarity (water, alcohol, or acetone) into an animal organism, cell culture, or model system containing membrane suspension was always accompanied by changes in the structural characteristics of the lipid bilayer of membranes [3-6]. This also applies to "apparent" concentrations, when the probability of at least one molecule of BAC to be present in the membrane suspension is close to zero.

Today, there exist numerous hypotheses attempting to account for this phenomenon of specific effects of BACs at ultra low concentrations on biological objects [7-10]. In our opinion, the main difficulty in substantiating such hypotheses is the explanation of the primary act of interaction between the molecules of the compound of interest with biological targets, when the ratio between the number of BAC molecules and the number of protein molecules is $1 : 10^4$ and lower. In view of this, many authors seek to account for the observed consistent patterns in terms of the structure-forming properties of polar (in particular, aqueous) solvents [11, 12].

The goal of this study was to determine the role of solvent polarity in the mechanism of action of BACs at ultra low concentrations. For this purpose, we performed a comparative study of the effect of the natural antioxidant α-tocopherol (α-TP), prepared in a polar solvent (ethanol-aqueous solutions with predomination of water upon diluting) and in a nonpolar solvent (based on petrolatum oil), within a broad concentration range (10^{-4} – 10^{-25} M), including the "apparent" concentrations (10^{-18} M and lower), on the viscosity properties of the lipid bilayer of mouse hepatocyte microsomal membranes *in vitro*.

Materials and Methods

Polar solutions of α-TP were obtained by the method of successive dilutions (by an order of magnitude) of the original α-TP solution (10^{-1} M) first with high-purity ethanol (Merck, Germany) to a concentration of 10^{-3} M and then with distilled water. Nonpolar solutions of α-TP were also obtained by the method of successive dilutions of the initial solution (10^{-1} M) with petrolatum oil, which is a side product of oil refining and represents a mixture of mineral oils containing solid paraffin hydrocarbons.

Hepatocyte endoplasmic reticulum membranes of F_1(C57 x DBA_2) mice were isolated by successive centrifugation as described in [13]. Before experiments, mice were decapitated. Protein concentration was determined by the method of Lowry [14].

The parameters characterizing the viscosity properties of membranes were calculated from the EPR spectra obtained at 293 K on a Bruker-200D spectrometer (Germany) [15]. The stable nitroxyl radicals 5-doxyl-stearic acid (C_5) and 16-doxylstearic acid (C_{16}) were used as spine probes to study the rigidity of surface (~8 Å) and microviscosity of deep (-20 Å) areas of lipid bilayer, respectively. Membrane rigidity in the area of location of the C_5 probe was described by the parameter ΔH_{II} (the distance between the low-field and high-field spectral components). This parameter was proportional to the ordering parameter *S*, which characterizes the lability of fatty-acid chains of lipids in the area of location of the probe. Membrane microviscosity in the area of location of the C_{16} probe was described by the rotational correlation times τ_{c1} and τ_{c2}, which were determined by the equations for rapid anisotropic rotation of radicals [15].

Data were statistically processed using the Statistica software package at a statistical reliability of 95%.

Results and Discussion

As a result of studying the effect of polar and non-polar α-TP solutions on the microviscosity of deep areas of membrane lipid bilayer at 293 K, we obtained typical for the action of BACs at ultra low doses multimodal concentration dependences of effects expressed in percentage relative to the control, which for τ_{cl} are shown in Figure la. The maxima and minima of effects corresponded to increases and decreases in the microviscosity of deep areas, respectively, with the values of the effects of ultra low and physiological concentrations being comparable for both types of solvents. The effects whose values did not exceed 2% were regarded insignificant for the probe used (the crosshatched region on Figure la).

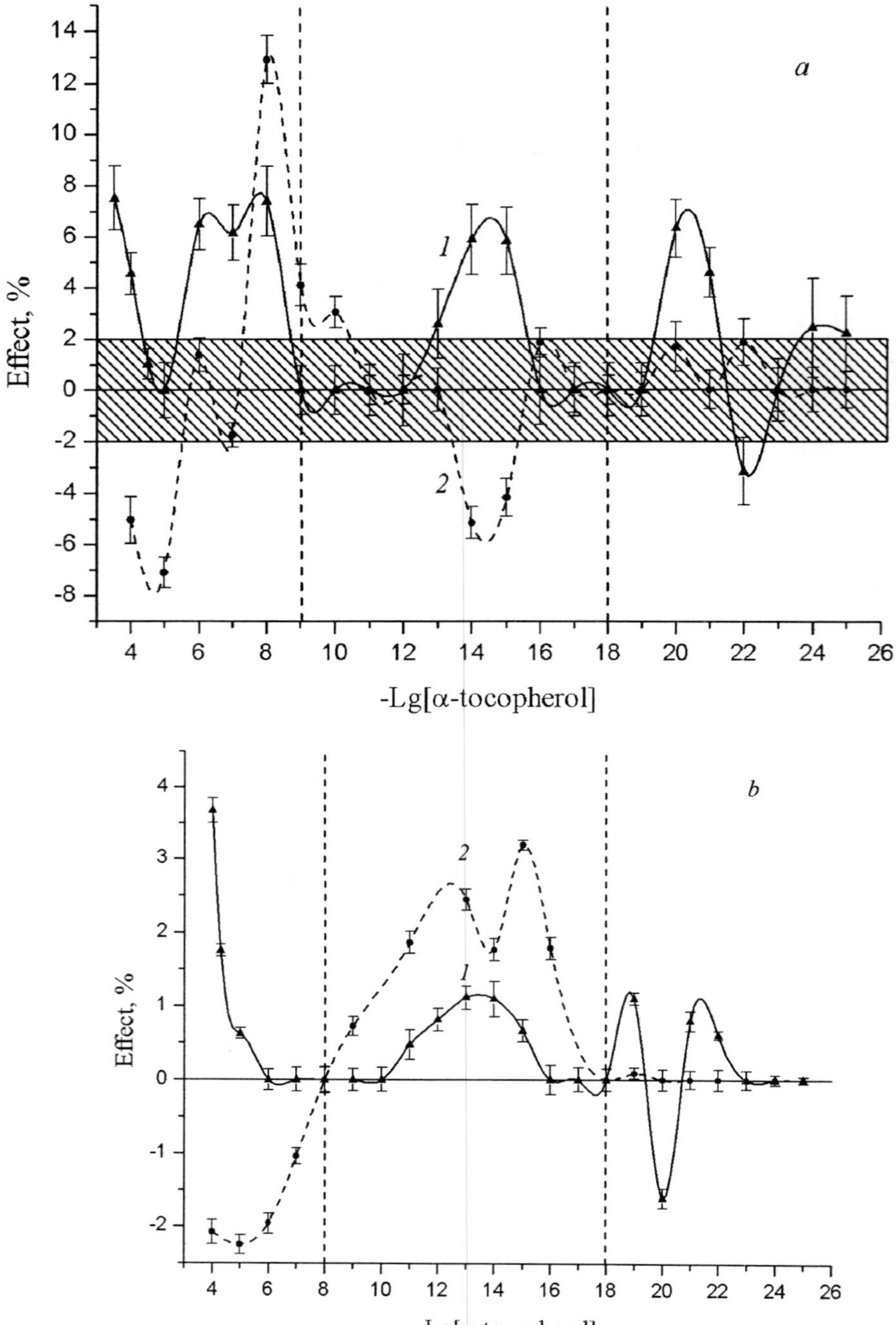

Figure. 1. Effect of α-tocopherol in (*1*) polar and (*2*) nonpolar solvents on the microviscosity of deep (a) and rigidity of surface (b) areas of microsomal membranes at 293 K. Protein concentration in the membrane suspension was 4 mg/ml.

Importantly, in the area of the so-called apparent concentrations (lower than 10^{-18} M), where the probability of at least one α-TP molecule to be present in solution was close to zero, a significant effect was exerted only by polar (in our case, aqueous) α-TP solutions (Figure la, curve *1*). This finding suggests that solvent polarity plays the key role in the transfer of information about a compound in one or another form upon dilution in the course of preparation of solutions.

We also observed the presence of the so-called dead areas on concentration curves, in which effects were absent. Note that the width of the dead zone decreased by two orders of magnitude when the solvent was changed for a nonpolar one (petrolatum oil) (Figure la, curve *2*).

Both curves shown in Figure la can be divided into three areas:

(1) 10^{-18}-10^{-25} M (the area of apparent concentrations, in which the effects of only α-TP solutions in the polar solvent were observed).
(2) 10^{-9}-10^{-18} M (the area of ultra low concentrations, in which the concentration-effect curves for α-TP in the polar and nonpolar solvents were antiphase but comparable in amplitude).
(3) 10^{-3}-10^{-9} M (the area of physiological concentrations, at which α-TP was effective in the organism; in this area, the concentration-effect curves for α-TP in the polar and nonpolar solvents were synphase, but the amplitude was greater in the case of α-TP solutions in petrolatum oil).

Figure lb shows the concentration dependences of the effects of polar and nonpolar α-TP solutions on the rigidity of surface areas (~8 Å) of the lipid bilayer of microsomal membranes at 293 K. Both concentration-effect curves also had a multimodal nature, when the maxima and minima of effects corresponded to increases and decreases in the lipid bilayer rigidity in the area of the C_5 probe location. As in the case of effects observed in deep membrane layers, similar concentration ranges could be distinguished on the concentration-effect curves obtained for the surface lipid layers of membranes. In the range of physiological concentrations of α-TP (10^{-4}-10^{-8} M), the concentration curves for both solvent types were antiphase, with larger amplitudes being recorded in the case of polar α-TP solutions. In the area of ultra low concentrations (10^{-8}-10^{-18} M), the curves were synphase, with larger amplitudes being recorded in the case of nonpolar (petrolatum oil-based) α-TP solutions. However, in the area of apparent concentrations (10^{-18}-10^{-25} M), a reliable and statistically significant effect was observed only for α-TP solutions in the polar (aqueous) solvent, which additionally confirmed the hypothesis on the key role of solvent polarity in the transfer of information on about a compound in one or another form upon dilution. Dead zones, in which no significant effect was detected, were also observed only in the case of the polar solvent.

CONCLUSIONS

Thus, this was the first work to study the role of solvent polarity by the example of the effect of α-TP within a broad concentration range (up to ultra low and apparent concentrations) on viscosity properties of mouse hepatocyte microsomal membranes *in vitro.*

In the area of physiological and ultra low α-TP concentrations, phase mode of the concentration-effect curves was observed both in the case of polar and nonpolar solvents. Phases were opposite in the surface and deep layers of the lipid bilayer of microsomal membranes. Upon the transition from the polar to nonpolar solvent, the dead zone either disappeared (in the case of C_5 probe) or narrowed (in the case of C_{16} probe). In the area of apparent α-TP concentrations ($<10^{-18}$ M), only polar (aqueous) α-TP solutions had a significant effect of the structure of microsomal membranes.

REFERENCES

[1] E.B. Burlakova, A.A. Konradov, and E.L. Mal'tseva, *Khim. Fiz.*, Vol. 22, N 2, pp. 21-40, 2003.

[2] E.B. Burlakova, *Ross. Khim. Zh.*, Vol. 43, N 5, pp. 3-11, 1999.

[3] N.P. Pal'mina, N.G. Bogdanova, E.L. Mal'tseva, and E.I. Pynzar', *Biol. Membr*, Vol. 9, N 8, pp. 810-820, 1992.

[4] V.E. Zhernovkov, N.G. Bogdanova, T.V. Lelekova, and N.P. Pal'mina, *Rad. Biol. Radioekol*, Vol. 43, N 3, pp. 331-333, 2003.

[5] E.L. Mal'tseva and N.P. Pal'mina, *Rad. Biol. Radioekol.*, Vol. 43, N 3, pp. 301-305, 2003.

[6] V.V. Belov, E.L. Mal'tseva, and N.P. Pal'mina, *Rad. Biol. Radioekol*, Vol. 43, N 3, pp. 306-309, 2003.

[7] A.L. Blyumenfel'd, *Biofizika*, N 1, p. 129, 1993.

[8] E.B. Burlakova, A.A. Konradov, and I.V. Khudyakov, *Izv. Akad. Nauk, Ser. Biol.*, N 2, pp. 184-193, 1990.

[9] S.V. Zaitsev, A.M. Efanov, and L.A. Sazanov, *Ross. Khim. Zh.*, Vol. 43, N 5, pp. 28-33, 1999.

[10] I.P. Ashmarin, E.P. Karazeeva, and T.V. Lelekova, *Ross. Khim. Zh.*, Vol. 43, N 5, pp. 21-28, 1999.

[11] V.I. Lobyshev, R.E. Shikhlinskaya, and B.D. Ryzhikov, *J. Mol. Liquids*, Vol. 82, p. 73, 1999.

[12] G.S. Anagnostatos "Experimental detection of the existence of structures in high dilutions", pp. 326-334, in the book *"High Dilution Effects on Cell and Integrated Systems"*, Ed. by C. Faddei-Ferretti, P. Marotta, Singapore, New-Jersey, London, Hong-Hong, World Scientific, 1999.

[13] K.Y. Hostetler, D.B. Zenner, and H.P. Morris, *Biochim. Biophys. Acta*, Vol. 441, pp. 231-238, 1976.

[14] O. Lowry, N. Rosenbrouch, A. Barr, and R. Randall, *J. Biol. Chem.*, Vol. 193, pp. 265-275, 1951.

[15] A.V. Kuznetsov, "Metod Spinovogo Zonda", Moscow, *Nauka,* 1976, 210 pp.

In: New Aspects of Biochemical Physics
Editors: Sergei D. Varfolomeev et al., pp. 23-26
ISBN 1-60021-419-3

Chapter 3

RESEARCH OF NO-INHIBITING AND ANTIRADICAL ACTIVITY OF NITROUS HETEROCYCLES β-HYDROXYDERIVATIVES (PYRIDINE AND BENZIMIDAZOLE)

T. N. Nosko, Yu. V. Kuznetsov and L. D. Smirnov
N. M. Emanuel Institute of Biochemical Physics,
Russian Academy of Sciences
4, Kosygin str., Moscow 119991, Russia

ABSTRACT

Using homogeneous hydrophilic chemiluminescent model system the antiradical activity of 3-hydroxypyridine (3-HP) and 5-hydroxybenzimidazole (5-HB) derivatives was studied. By introducing of lipopolysaccharide endotoxin NO-inhibiting activity of 3-HP and 5-HB derivatives in mice liver was researched. It was estimated that some substances had antiradical activity and could changed NO synthesis in mice liver. One of them, the 3-HP derivative (4-methylthyobenzimidazolile), had highest antiradical and NO-inhibiting activity.

Key-words: 3-hydroxypyridine and 5-hydroxybenzimidazole derivatives, NO-inhibiting and antiradical activity.

1. INTRODUCTION

In pathogenesis of many diseases is observed disbalance for the benefit of fermentative and inorganic synthesis of the active forms of oxygen and nitrogen, including peroxide of hydrogen, super oxide anion-radical, hydroxyl radical, singlet oxygen in comparison with activity of system redox-homeostase. The exclusive role in these pathological processes is played by a two-nuclear radical - NO, the opening of which biological properties was marked by the Nobel Prize [1-2]. In this connection intensive search of new compounds with NO-modulating activity now is conducted with the purpose of creation of preparations for treatment neurodegenerative and infectious diseases, various kinds of a shock, consequences insults, heart attack etc.

Among known synthetic antioxidants the significant interest is caused derivatives of 3-HP, close on the structure to group of vitamins B6 [3, 4] and derivatives of 5- HB, of a fragment, included in structure, of vitamin B12 [5]. The specified compounds inhibiter reactions free radical oxidation (FRO), change a structurally functional condition of membranes, receptors and membraneconnected ferments etc. [6, 7]. On the basis of derivatives of 3-HP some effective domestic medicinal preparations emoxipin and mexidol are created. The regulator of growth of plants ambiol is derivatives of 5-HB. Stated opens ample opportunities for development of new biologically active substances in the specified numbers of compounds [3-5].

The purpose of work - research NO-modulating and ARA of derivatives of 3-HP and 5-HB.

2. Technique of Research

Influence derivatives of 3-HP and 5-HB on synthesis NO in a liver white outbred of mice mails of weight 25-30 was estimated. To mice intraperitoneal, (i/p) for 4 h up to death lethal injection entered lipopolysaccharide endotoxin E. coli in a doze of 1,5 mg/kg. The researched substances by an animal (n=7) entered i/p [0,5 ml in isotonic solution chloride sodium (NaCl)] for 1 h up to death lethal injection. In 0,5 h after introduction of substances entered " a spin trap ". Intact animals (n=7) received isotonic solution NaCl in the same quantity by a similar way. The contents NO in samples of a liver determined by the modified method A.F. Vanina etc. [2].

ARA of derivatives of 3-HP and 5-HB was studied with the help hydrophilic chemiluminescent model system of oxidation consisting from hemoglobin, peroxyde of hydrogen and luminol [8]. In quality investigated parameters served the maximal intensity of a luminescence [amplitude hemiluminescence (HL)] and time from the moment of introduction of the initiator (H_2O_2) to the beginning development of a luminescence (latent period). Into structure of reactionary medium entered: 50 mM K^+-phosphorous buffer (pH 7,4), 2 mcM hemoglobine, 100 mcM luminol, 100 mcM ETDA, and also derivatives of 3-HP and 5-HB in various concentration. The reaction was begun by addition 100 mcM H_2O_2. Kinetics HL registered on spectrofluoriphotometer ("Shimadzu", Japan). For a quantitative estimation of ability derivatives of 3-HP and 5-HB to react with radicals located in a water phase of the given model system, the results of suppression HL recalculated in coordinates of the equation Shtern-Folmera [9]:

$$A_0 / A_i = 1 + K_i C,$$

Where A_0 and A_i are amplitude HL of model system without and at presence inhibitor; Ki is a constant of suppression HL, conditionally equal to a degree of interception inhibitor radicals formed in model system; C - molar concentration inhibitor.

3. Results of Research

The data on study ARA of derivatives of 3-HP and 5-HB with the help homogeneous hydrophilic chemiluminescent model system are submitted in tab. 1.

It is visible from it, all investigated substances in a various degree inhibited development HL, inducted by reactions FRO. On ARA they can be located in the following sequence: VIII VI > VII > II, III > IV > I > V. On a background of action endotoxin it was revealed, that among derivatives of 3-HP and 5-HB are available as inhibitors of production NO, and it stimulators (tab. 2).

Thus essential influence on character modification rendered anti-ions. In particular, the basis (III) essentially did not influence on the contents NO, and derivative of 3-HP (IV) in the form hemisuccinate in 1,5 times ($p< 0,05$) increased the contents NO. Among investigated derivatives of 5-HP similar effect also was found out. The salt (VI) with oxalic acid did not change production NO, while 5-HB glutamate (V) in 1,6 times ($p< 0,05$) reduced the contents NO.

Table 1. Constants of suppression (K_i) hemiluminescence β-hydroxyderivatives nitrous heterocycles (derivatives of 3-HP и 5-HB)

№	Compound	$K_{i,}$ M^{-1}
I	2-ethyl-6-methyl-3-HP	2.4×10^4
II	N-acetylcisteinat 3-HP	1.25×10^5
III	2,4,6-trimethyl-3-HP	1.25×10^5
IV	3-HP hemisuccinate	2.7×10^4
V	5-HB glutamate	1.5×10^4
VI	5-HB oxalate	3.0×10^5
VII	5-HB orotate	2.0×10^5
VIII	4-methylthyobenzimidazolile-3-HP	4.5×10^5

Table 2. Influence β-hydroxyderivatives nitrous heterocycles (derivatives of 3-HP и 5-HB) on NO synthesis in mice liver

№	Compound	Inhibition, %
I	2-ethyl-6-methyl-3-HP	66
II	N-acetylcisteinat 3-HP	115
III	2,4,6-trimethyl-3-HP	112
IV	3-HP hemisuccinate	155*
V	5-HB glutamate	61*
VI	5-HB oxalate	117
VII	5-HB orotate	122
VIII	4-methylthyobenzimidazolile-3-HP	13*

Note * – differences statistically significant in comparison with control ($p<0,05$), taken for 100%

So, generalizing the received results, it is possible to conclude, that among the tested substances there are compounds with ARA, and also varying synthesis NO in a liver of mice. Thus the greatest attention pays on itself derivative of 3-HP (4-methylthyobenzimidazolile -3-HP, VIII), having the highest ARA and ability inhibit synthesis NO in a liver of mice.

REFERENCES

[1] Moncada S. // *Acta Cardiol.* - 2004. - Vol. 59. - Suppl. 1: 47 50.

[2] Vanin A.F., Mordvincev P.I., Kleschev A.L. // *Studia Biophys.* -1984.- Vol. 102.- N 2. pp.135-143.

[3] Smirnov L.D. // Chemical and biological kinetics. New horizons. - M, 2005. – *"Chemistry"*, pp 102-129.

[4] Smirnov L.D., Dumaev K.M. // *Science to manufacture*. 2002.- №3.-pp. 53-55.

[5] Smirnov L.D., Kuznetsov Y.V., Stolyarova L.G., Lezina V.P. // *AS USSR, "Chem"* – 1985.-№ 8.- pp. 1855-1858.

[6] Voronina T.A. // *Psyhofarmac. biol. narc* - 2001. - T. 1, №1. - pp. 2-12.

[7] Dumaev K.M., Voronina T.A., Smirnov L.D. Antioxidants in preventive maintenance and therapy of pathologies CNS. - M.: . *Institute of biomedical chemistry RAMS*, 1995. pp.272.

[8] Klebanov G.I., Teselkin Y.O., Babenkova I.V. etc. // *The bulletin RAMS*. -1999. - № 2. – pp. 15-22.

[9] Teselkin Y.O., Babenkova I.V, Lyubicky O.B. etc. // *Questions of medical chem.* – 1997.- T. 43. - pp. 87-92.

In: New Aspects of Biochemical Physics
Editors: Sergei D. Varfolomeev et al., pp. 27-31
ISBN 1-60021-419-3

Chapter 4

PHASE SPACE RESEARCH ALGORITHM OF REACTIVE DISTILLATION PROCESSES

A. M. Shchekin, S. D. Tishaeva, Yu. A. Pisarenko and L. A. Serafimov
Moscow State Academy of Fine Chemical Technology named after Lomonosov
86, Vernadsky str., 119581 Moscow, Russia

ABSTRACT

The method of research for phase diagram structures of reactive distillation process in systems with many chemical reactions is offered. Use of the modified variables offered by Zharov for a finding of special points on chemical reactive manifold and their types is entered.

Key-words: CRM – chemical reactive manifold, RD – reactive distillation.

In connection with creation without waste manufactures special importance is got with development of new technological principles and receptions in technological manufactures. Creation of manufactures of chemical products on the basis of a principle of overlapping of chemical transformation and mass transfer – one of methods of the decision of the problems facing to a chemical industry.

As a whole advantages of the combined processes play a special role in development of waste technologies of reception of various organic synthesis products. Here special value gets increase in selectivity of processes and degrees of conversion of reagents for one pass of raw material through the device. The first leads to decrease in quantity of by-products, that in turn in full or in part removes questions of their recycling or destruction. The second creates necessary conditions for decrease in number and volume external recycles. Reagent losses are as a result reduced, and the degree of their use increases.

One of the first stages of development of the combined continuous RD processes is the analysis of their statics [1], allowing to determine basic opportunities of reactive mass transfer process. Results of the analysis of a statics of the combined process make a basis of its modelling and calculation as it allows to receive optimum variants of the organization of the combined process.

The basic role which is allocated in the analysis of a statics to structures of RD diagrams, causes a urgency of studying of local and not local laws as in the set it defines the order of stacking in phase space of curve beams of reactive distillation, i.e. a phase portrait of process. Besides a phase portrait defining influence on structure of grocery streams is rendered also with a configuration of phase space. The knowledge of type of the phase diagram allows to allocate modes of carrying out of reactive distillation which provides the highest output of a target product and an opportunity of allocation from a reactionary mix of substances making it.

It is known, that in case of equilibrium distillation with instant chemical reactions, the phase space of corresponding dynamic system represents chemical balance multiple (CRM) on which curves of process in the form of beams of RD lines are located.

Research of phase space properties of equilibrium distillation process with chemical reactions carry out by means of construction of the grid located in phase space. It is above noted, that in this case the phase space represents curvilinear CRM, direct construction on which grid in regular intervals covering it inconveniently.

The basic idea of work consists in constructing the specified grid on boundary linear multiples of a concentration simplex and then to design it on chemical balance multiples along chemical interaction multiples. Considering, that crossing of the given multiples – chemical interaction multiples and chemical balance multiples – has dimension zero, we shall receive, that to each unit of the grid constructed on boundary multiple, the unit of the grid constructed on CBM of the same dimension (Figure 1).

Describe one of possible variants of specified grid construction, based on construction of sets of secants and sections on the certain, in advance chosen boundary surfaces of a concentration simplex. Remind, that the secant a surface is represented by structures with the fixed relation of concentration of any pair substances while the section is formed by structures with constant concentration of one of them. Grid construction we shall consider on an example of a side of the concentration simplex dimensional h – 1 on which it is presented h substances. In this case each secant is linear multiple (hyperplane) with dimension (h – 2). Being crossed among themselves, the given multiples form units of a grid.

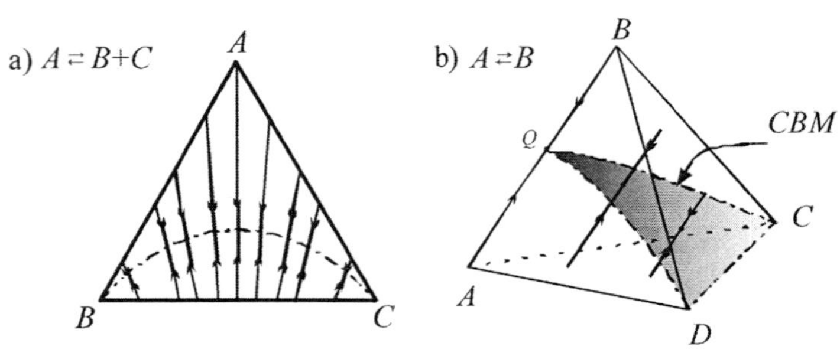

Figure. 1. Interrelation between boundary multiples of a concentration simplex and CRM: –·–·–·– CRM; ⟶ a chemical interaction line.

Note, that sets of sections and secants include boundary elements of a considered side. As along any boundary element of a side concentration, at least, one of the substances, presented on it, is equal to zero (and, hence, it is constant), the given boundary element belongs to section with zero concentration of the specified substance. And itself (h – 1) the dimensional side is included in section with the concentration of substance absent on it fixed and equal to zero.

Similarly each of the secants determined by the constant relation of concentration of pair substances, present on the verge, includes as limiting elements all borders of the specified side where simultaneously there are no both these substances, making the considered pair. Besides it partially includes also borders on which simultaneously they are present. We shall lead numbering substances (u = 1.., h) on all separately the boundary varieties belonging an initial surface. On each of (h – 1)-dimensional sides the substance with maximal number u_{Max} = h is allocated, and concerning the given component the sections belonging considered boundary multiple (Figure 2) are spent.

Special points localization of of phase space can be lead as directly by an estimation of the rate formed of the right parts of dynamic system in RD process, and isotherm multiple construction which behaviour in a vicinity of a special point defines its type according to Schreinemakers theorem.

For the first time the thermodynamic analysis of equilibrium distillation process at presence of chemical balance is lead by V.T.Zharov [2]. By him it has been established, that besides usual special points in system are present so-called reactive azeotrope – conditions in which there is indemnification of changes of the structure caused by processes of evaporation and chemical reaction. Thus for presence reactive azeotrope a necessary and sufficient condition is the arrangement liquid-vapor line on chemical interaction multiple [3].

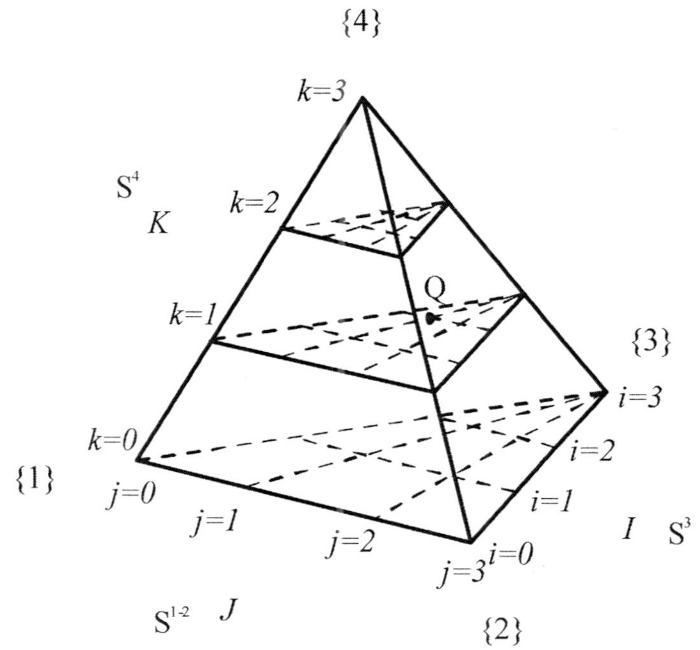

Figure. 2. An arrangement of sections on boundary multiple with dimension 3 of concentration simplexes of higher dimension.

Enter the modified variables for systems with instant chemical reactions [2]:

$$\alpha_k^{(1)} = w_k^{(1)} - \sum_{s=1}^{r} w_s^{(1)} \sum_{\rho=1}^{r} \gamma_{i,r} \frac{A_{i,r}}{\det[\mathbf{B}_K]},$$
$$\alpha_k^{(2)} = w_k^{(2)} - \sum_{s=1}^{r} w_s^{(2)} \sum_{\rho=1}^{r} \gamma_{i,r} \frac{A_{i,r}}{\det[\mathbf{B}_K]}, \quad (1)$$

Where $\alpha_k^{(1)}, \alpha_k^{(2)}$ – the transformed variables of structure for a vapor and liquid phase accordingly, i = r+1, r+2, …, c-1, $A_{i,r}$ – algebraic addition of an element $\gamma_{i,r}$ matrixes$[B_K]$, w_k – a mass fraction of substance, thus the index (1) concerns to a liquid phase, and (2) - to vapor; r - number of linear independent reactions, c – quantity of substances i, $\gamma_{i,r}=\nu_i M_i$ – multiplication of stoichometric coefficient of substance R_i on it molar weight. And $[B_K]$ is a matrix modified stoichometric coefficients of the size $r \times r$ made of the first elements of a matrix [B] where its determinant is not equal to zero:

$$[\mathbf{B}] = \begin{bmatrix} \mathbf{B}_K \\ \cdots \\ \mathbf{B}_O \end{bmatrix} = \begin{bmatrix} \gamma_{1,1} & \gamma_{1,2} & \cdots & \gamma_{1,r} \\ \gamma_{2,1} & \gamma_{2,2} & \cdots & \gamma_{2,r} \\ \cdot & \cdot & \cdot & \cdot \\ \gamma_{r,1} & \gamma_{r,2} & \cdots & \gamma_{r,r} \\ - & - & - & - \\ \gamma_{r+1,1} & \gamma_{r+1,2} & \cdots & \gamma_{r+1,r} \\ \cdot & \cdot & \cdot & \cdot \\ \gamma_{c-1,1} & \gamma_{c-1,2} & \cdots & \gamma_{c-1,r} \end{bmatrix} \begin{matrix} [\mathbf{B}_K] \\ \\ [\mathbf{B}_O] \end{matrix} . \quad (2)$$

Introduction of the modified variables allows to receive in this case dynamic system of externally similar process of the open evaporation:

$$\frac{d\alpha_k^{(1)}}{d\ln m} = \alpha_k^{(2)} - \alpha_k^{(1)}, k = r+1, r+2, ..., c-1, \quad (3)$$

where m – solution weight.

It is necessary to note, that new variables $(\alpha_k^{(1)}, \alpha_k^{(2)})$ can be both positive, and negative. For the specified set of the modified variables the system of phase space coordinates is realized.

Special points of equation system (3) are in which

$$\alpha_k^{(2)} - \alpha_k^{(1)} = 0, \quad k = r+1, r+2, ..., c-1. \quad (4)$$

It is visible, that special points, as well as earlier, there will be the points of concentration space corresponding components and azeotrpes, i.e. points in which structures of a liquid and vapor phase are identical, and also reactive azeotropes (in which structures of a liquid and vapor are various).

The modified variables together with a condition (4) are used for the analysis of phase space of dynamic system of reactive distillation.

The method of the analysis of phase space in RD process in conditions of chemical balance has been used for synthesis of possible variants of division schemes of some industrial reactive systems, such as reception mesithyl oxide, ethil acetate, methyl tert-butyl ether, diisopropyl ether, etc.

REFERENCES

[1] M.I Balashov, L.A Serafimov. Analysis of a continuous processes statics of combined reactive distillation // *Theor. Bases of Chem.Technol.* - 1980. – (14). - №6. - pp. 803 - 808.

[2] V.T. Zharov. Heats research of the open phase processes and not local laws of phase balance diagrams in various heterogeneous systems: *PhD Dissertation*. – Leningrad., 1969. – pp.474.

[3] Yu.A Pisarenko, L.A Serafimov. Stationary conditions of distillation process with chemical reaction // *Theor. Bases of Chem.Technol.* 1991.-(25). - №3.-pp. 422-426.

In: New Aspects of Biochemical Physics
Editors: Sergei D. Varfolomeev et al., pp. 33-40
ISBN 1-60021-419-3

Chapter 5

DEFINITION OF OPTIMAL OPERATING MODES OF DICHLOROACETIC ACID HYDROGENATION REACTOR

A. G. Loschev**, *E. A. Glushachenkova and Yu. A. Pisarenko
M.V.Lomonosov Moscow State Academy of Fine Chemical Technology,
pr. Vernadskogo, 86, Moscow, 119571, Russia
Department of Chemistry and Chemical Engineering for Basic Organic Synthesis

ABSTRACT

There have been defined the optimal operating modes of the column fixed-bed catalyst reactor of dichloroacetic acid (DCAA) hydrogenation, providing both the high purity of commercial monochloroacetic acid (MCAA) and rational using of the expensive palladium activated charcoal-based catalyst. The most rational methods of sectioning and additional wetting of the reactor have been proposed.

Key words: Monochloroacetic acid, dichloroacetic acid, hydrogenation of dichloroacetic acid, palladium activated charcoal-based catalyst, processing of dichloroacetic acid, column fixed-bed catalyst reactor, optimal operating modes.

NOMENCLATURE

C_{DCAA}	dichloroacetic acid (DCAA) concentration, mol/l
C_{H2}	hydrogen concentration, mol/l
C_{MCAA}	monochloroacetic acid (MCAA) concentration, mol/l
D	reactor diameter, mm
$E_{a,1}$	activation energy of reaction (1) is equal to 62.85 kJ/mol
$E_{a,2}$	activation energy of reaction (2) is equal to 65.20 kJ/mol

* Tel.: 7 (495) 936-8905, fax: 7 (495) 434-8711; e-mail: diplom_artem@mail.ru

F	intermediate stream rate, in kg/h or in % of the reactor product stream
h	reactor height, m
h_1	first section height, m
h_2	second section height
$k_{0,1}$	pre-exponential factor for reaction (1), is equal to $8.9377 \cdot 10^7$ l/l·h
$k_{0,2}$	pre-exponential factor for reaction (2) is equal to $2.2344 \cdot 10^7$ l/l·h
r_1	DCAA hydrogenation rate (reaction (1), mol/l·h
r_2	MCAA hydrogenation rate (reaction (2), mol/l·h
R	universal gas constant is equal to 8.3144 J/mol·K
T_F	temperature of intermediate stream, °C
T	temperature of reaction (1) and (2), K
t	time, h.

INTRODUCTION

Monochloroacetic acid (MCAA) relates to the most important products of industrial organic synthesis: it's a raw for the production of carboxymethylcellulose, herbicides and chelators and also applied in aniline-dye, pharmaceutical and other industrial fields. The world MCAA manufacturing capacity was equal to 240 thousand tons in 1990, and had reached a level of ~ 500 thousand tons by 2000 [1].

Industrial methods of the MCAA synthesis are sulfuric acid hydration of trichloroethylene or catalytic chlorination of acetic acid (AA). The first of them is characterized by severe synthesis conditions, complicated equipment and great waste bulk and is out-of-date, so the second way takes the leading position in the world balance of MCAA production [1].

When obtaining MCAA by catalytic chlorination of AA, up to 8.0 wt. % of the more chlorinated by-product, dichloroacetic acid (DCAA) is formed. At the same time it makes sufficiently rigid requirements to the commercial MCAA purity – the DCAA concentration mustn't be more then 0.1–0.2 wt. % in it, especially if MCAA is intended for the production of cellulose esters or herbal protectors. Therefore, an MCAA synthesis flowsheet usually includes a DCAA processing stage.

Now several ways of purifying of the MCAA synthesis reactionary mixture from DCAA are known. Among them it's worth noting the method of MCAA melt crystallization [2], liquid-phase catalytic DCAA hydrogenation at the presence of MCAA natrium salts and the palladium activated charcoal-based catalyst [3], azeotropic or extractive distillation [4]. Another method of purifying of the reactionary mixture from DCAA is its treatment by ozone or a water solution of hydrogen peroxide at a temperature of 70.0–120.0° C [2]. Only two ways have been widely adopted in the industry: the MCAA melt crystallization and liquid-phase catalytic DCAA hydrogenation on the catalyst of palladium deposited on activated charcoal (without MCAA natrium salts) [5].

The most effective catalysts of the DCAA hydrogenation are noble metals of the *VIII* group of Periodic system, but in the industry palladium is preferred [6]. At that a range of supporters applied isn't so wide: this is silica gel, Al_2O_3 or kieselguhr, and also activated charcoal, the latter is the most preferable one, because the others dissolve in the reactionary medium.

In the industry the DCAA catalytic hydrogenation to MCAA is carried out on the catalyst containing 1-2 wt. % Pd deposed on activated charcoal, at an eightfold excess of hydrogen in relation to the DCAA concentration in the feed mixture (to prevent the catalyst deactivation) and at a temperature of 130.0–145.0° C. Since activated charcoal is sensitive to abrasion, the process is realized in a column fixed-bed catalyst reactor, at that both gaseous hydrogen and the liquid-phase mixture containing DCAA are introduced top-down in it (by cocurrent) [7,8].

In the work [8] it has been shown, that the DCAA hydrogenation can be effectively carried out solely in case of the liquid being distributed entirely and evenly on the catalyst surface as a film in the reactor, i. e. at an unit-area liquid flow rate of 0.30–1.55 m3/m2·h. A liquid bulk reduction results in overheating of the catalyst and its subsequent resinification, and increasing makes difficulties for hydrogen access to the catalyst surface, both lead to the catalyst deactivation and reaction rate lowering.

A temperature in the reactor makes an essential impact on an unit-area liquid flow rate value: the DCAA hydrogenation passes at a temperature above 125.0–130.0° C, yet its exceeding of 145.0° C isn't reasonable because of a liquid bulk decrease. The problem of maintaining of the adjusted process temperature, and consequently, the unit-area liquid flow rate is complicated by the fact, that the reactions

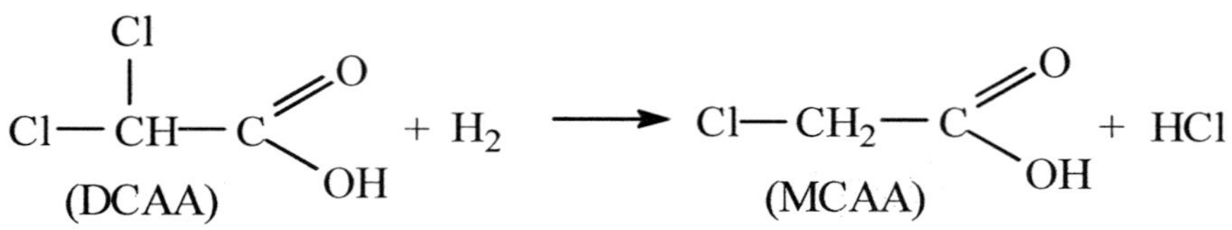

(1)

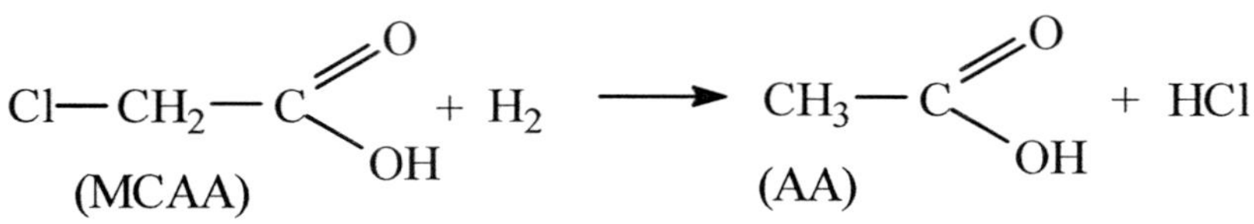

(2)

taking place during the DCAA hydrogenation are strongly exothermal ones: their heat efficiencies (under standard conditions) are equal to –90.6 and –92.0 kJ/mol accordingly.

The mentioned above factors cause a necessity of an optimum operating conditions choice of the DCAA hydrogenation reactor, providing both the high purity of commercial MCAA and rational using of the expensive palladium catalyst.

Now against the background of increasing of demand for high-purity MCAA and palladium catalyst costs this problem has been becoming more and more actual, both in case of the new plant design and modernization of the existing. In the latter case, due to a change of the raw composition or capacity as compared with designed values, there appears the task of an optimum operating mode choice for the reactor with already known constructional parameters. The present work is devoted to deciding of such a problem.

EXPERIMENTAL

As an object of optimization there has been considered the two-sectional DCAA hydrogenation column reactor with a diameter D of 540 mm and a height h of 12.0 m, the height of the first section h_1 is 5.0 m, and of the second h_2 – 7.0 m. In the given reactor an unit-area liquid flow rate is maintained at a level of 0.30–1.55 $m^3/m^2 \cdot h$, a temperature is 130.0–145.0° C, a DCAA conversion is above 77.0%, and a MCAA conversion less than 10.0% at a raw charge of 960 kg/h and a hydrogen flow rate of 27.7 kg/h (a H_2 - DCAA weight ratio is equal to 8 : 1). At that from 23.5 to 35.0% of the reactor feed charge is directed to the second section entrance. The feed has the following composition (in wt. %): HCl – 0.9, H_2O – 0.9, AA – 28.1, MCAA – 47.1, DCAA – 23.0.

In the present work there has been investigated an opportunity of maintaining of a temperature at a level of 130.0–145.0° C, unit-area liquid flow rate within the limits of 0.30–1.55 $m^3/m^2 \cdot h$ and a DCAA conversion – above 77.0% in the reactor with the same constructional parameters, the feed composition and a weight H_2 - DCAA ratio, but for a lower feed charge of 720 kg/h and hydrogen flow rate of 20.8 kg/h. Since MCAA is the target product, its conversion mustn't exceed 10.0%.

The estimation has been fulfilled on the base of the proprietary software of *PRO/II* with use of the mathematical model of an adiabatic plug flow reactor. When calculating the vapor-liquid equilibrium there has been used the equation of *NRTL* [9] for liquid activity coefficients, vapor fugacity coefficients have been determined by the Hayden-O'Connell method [10], and the solubility of the incondensable (under reaction conditions) components of H_2 and HCl – by Henry's law.

The rates of the reactions (1) r_1 and (2) r_2 were calculated by the equations:

$$r_1 = \frac{dC_{DCAA}}{dt} = -k_{0,1} e^{\frac{-E_{a,1}}{RT}} \left[C_{DCAA} \right] \tag{3}$$

$$r_2 = \frac{dC_{MCAA}}{dt} = -k_{0,2} e^{\frac{-E_{a,2}}{RT}} \left[C_{MCAA} \right] \tag{4}$$

where C_{DCAA} and C_{MCAA} – concentrations of DCAA and MCAA in the liquid phase, mol/l, R – universal gas constant is equal to 8.3144 J/mol·K; T – temperature, K; t – time, h.

As the DCAA hydrogenation is implemented in the liquid phase, the hydrogen concentration is determined by its solubility, which is nearly independent on a process temperature [1]. So it has been accepted that $C_{H2}=const$ in the equations (3) and (4), at that

$$k_{0,1} = k'_{0,1} \left[C_{H_2} \right] \tag{5}$$

$$k_{0,2} = k'_{0,2} \left[C_{H_2} \right] \tag{6}$$

pre-exponential factors $k_{0,1}$ and $k_{0,2}$ are equal to $8.9377 \cdot 10^7$ and $2.2344 \cdot 10^7$ l/l·h, and activation energies $E_{a,1}$ and $E_{a,2}$ – 62.85 and 65.20 kJ/mol respectively. In accordance with the experimental data [8], a rate value of the reaction (2) doesn't exceed 25,0% of a rate of the reaction (1).

RESULTS

The calculations have shown, that at a feed charge of 720 kg/h a temperature along the reactor is within the limits of 130.0–145.0° C, and an unit-area liquid flow rate at the top of the catalytic bed is equal to 1.5 $m^3/m^2 \cdot h$. However, in the lower catalytic beds a lack of the liquid has been observed: an unit-area liquid flow rate takes on values below 0.30 $m^3/m^2 \cdot h$ (Figure 1) and in the bottom it comes to 0.20 $m^3/m^2 \cdot h$.

As it has been marked above, insufficient wetting of the catalyst surface results in decreasing of reaction rates. To decide the problem accrued, additional techniques have been used in the present work: sectioning of the reactor and imputing of raw MCAA, containing (in wt. %): AA – 4.0; MCAA – 90.0 and DCAA – 6.0, a portion of the reactor product stream (recycling) or an additional amount of AA (above contained in the raw) into its second section. At that there have been defined the height of the first section h_1, i. e. actually, the location of imputing of the intermediate stream (AA, raw MCAA of recycle), and also its rate F and temperature T_F. The results are presented in Tab. 1.

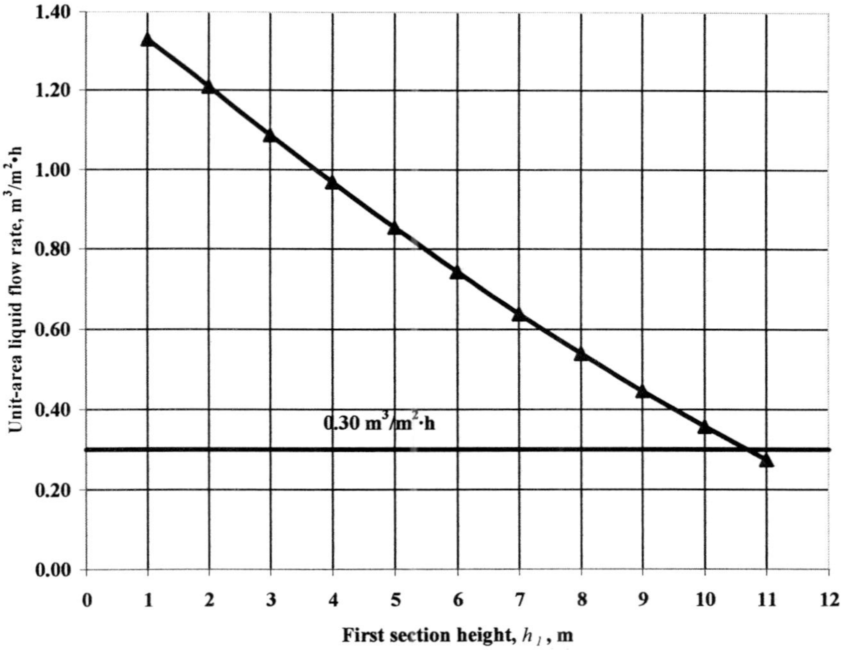

Figure 1. Unit-area liquid flow rate profile along DCAA hydrogenation reactor at feed charge of 720 kg/h

Table 1. Variants of sectioning of the DCAA hydrogenation reactor for various types of an intermediate stream inputted into the second section

T_F, °C	Reactor parameters	Type of an intermediate stream inputted into the second section		
		recycle	AA	raw MCAA
20.0	h_1, m	7.0–11.0	5.0–7.0	—
	F, kg/h	10.0–20.0*	50.0–110.0	—
70.0	h_1, m	—	—	5.0–11.0
	F, kg/h	—	—	70.0–90.0
130.0	h_1, m	8.0–9.0	5.0–9.0	7.0–9.0
	F, kg/h	2.5–15.0*	30.0–130.0	30.0–70.0

* - for the recycle a range of a value F is specified in % of a reactor product stream rate evaluated in kg/h

The investigation carried out has shown that at a feed charge of 720 kg/h the following way of sectioning of the DCAA hydrogenation reactor is the most reasonable one: the height of the first section h_1 is 7.0 m, of the second one h_2 is 5.0 m at the summary height h of 12.0 m. At that to keep up a temperature at a level of 130.0–145.0° C and an unit-area liquid flow rate within the limits of 0.30–1.55 $m^3/m^2 \cdot h$, it's possible to apply the below listed variants of inputting of the intermediate stream into the second reactor section:

- 50.0–110.0 kg/h AA at 20.0° C;
- 30.0–130.0 kg/h AA at 130.0° C;
- recirculation of 10.0–20.0% of the product stream after its cooling to 20.0° C
- and subsequent removing of the vapor phase;
- 70.0–90.0 kg/h raw MCAA at 70.0° C;
- 30.0–70.0 kg/h raw MCAA at 130.0° C.

In all the cases listed above a DCAA conversion is more than 77.0%, and even at the minimum intermediate stream rate (for the limits of values pointed in Tab.1) an unit-area liquid flow rate margin provided is more than 5.0%. However, a MCAA conversion takes on acceptable values merely when inputting the recycle or AA into the second section, and exceeds 30.0% when using raw MCAA.

Among of the above mentioned variants of additional wetting, the second is the best one, because it doesn't only provides specified values of a temperature, unit-area liquid flow rate and conversion of MCAA and DCAA, but permits to vary F within the wider interval in comparison with all the others, that is especially important when it's difficult to guarantee the constant flow rate.

The other ways of sectioning of the reactor (Tab. 1) are less preferable both due to some reduction of a unit-area liquid flow rate margin and a number of reserve variants of intermediate stream inputting. For example, when using the same way as at a feed charge of 960 kg/h, i. e. h_1 is 5.0 m and h_2 is 7.0 m, only three variants are possible:

- 50.0–110.0 kg/h AA at 20.0° C;
- 30.0–130.0 kg/h AA at 130.0° C;

- 70.0–90.0 kg/h raw MCAA at 70.0° C.

At that the variant of raw MCAA using is also unacceptable, so in this case a MCAA conversion exceeds 10.0%, and the best way is the second one permitting to change F within the larger interval.

The original flowsheet of the DCAA hydrogenation in the column reactor for a feed charge of 960 kg/h is presented in Figure 2a, and in Figure 2b and 2c – schemes chosen on the base of results obtained for feed charges of 720 and 960 kg/h.

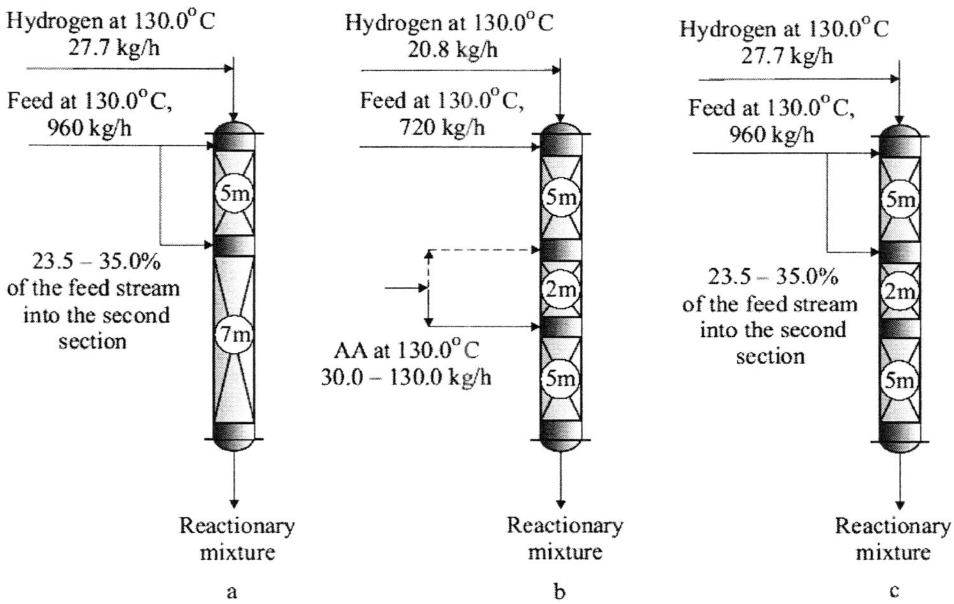

Figure. 2. Proposed variants of DCAA hydrogenation flowsheet in column reactor for different feed charges: a – original variant for a feed charge of 960 kg/h; b – 720 kg/h, less preferable variant of additional wetting is indicated with the dotted line; c – new variant for a feed charge of 960 kg/h

CONCLUSIONS

Consequently, to use the reactor with the same design characteristics (h = 12.0 m; D = 540 mm) at a feed charge of 720 kg/h as those of 960 kg/h, and at that to provide the optimal process temperature (130.0–145.0° C), unit-area liquid flow rate (0.30–1.55 $m^3/m^2 \cdot h$), conversion of DCAA (above 77.0%) and MCAA (less than 10.0%), it's necessary to apply the following techniques:

- additional sectioning of the available reactor, that doesn't suppose the sufficient modification of its construction imagined in Figure 2a, and, in fact, consists in dividing of the second section with a height of 7.0 m into two parts, i. e. the reactor must be composed of three series sections with heights of 5.0, 2.0 and 5.0 m;
- inputting an additional amount of AA (30.0–130.0 kg/h) at 130.0° C into the second or third section.

In the absence of an enough quantity of AA at a plant, it's possible to use alternative variants of supplementary wetting, supposing the input of the recycle into the second or third reactor section.

The operating modes selected for a feed charge of 720 kg/h are optimal ones in the view of temperature and unit-area liquid flow rate values and supply the required values of a DCAA and MCAA conversion, and also the steady reactor operation and the stable product mixture composition.

The sectioning method suggested for the reactor (Figure 2b and 2c) permits to manage the process flexibly and operatively enough at the componentwise variation of the feed composition within the limits of ±5.0 wt. % (of the appropriate component concentration in the feed mixture at the constant concentration ratio of the others) and changing of a feed charge within the bounds of 720–960 kg/h.

REFERENCES

[1] Zanaveskin, L.N., Bulanov, V.N., Kunitsyn, D.G., Uskach, Ya.L. and Varshaver, E.V., Production of monochloroacetic acid. Report 1: Processing of subsidiary dichloroacetic acid by technique of catalytic hydrogenation, *Catalysis in Industry, No.* 3, pp. 26–37, 2003.

[2] Smirnova, S.A. and Krasheninnikova, A.A., *Kislorodsoderjashie organicheskie soedinenuya* (Organic oxycompounds)(Khimiya Moscow), 1987, 39 p. *(in Russian)*

[3] Morozov, A.V., Fomin, V.A. and Kudryavtseva, T.G., Method for purification of monochloroacetic acid, *USSR Certificate (Patent) No.* 1 004 345, March 15, 1983.

[4] Horsley, L.H., Purification of 2-chloroalkanoic acids by azeotropic distillation, *USA Patent No.* 3 772 157, November 13, 1973.

[5] Rucker, J.T. and Sconce, J.S., Method for reducing dichloroacetic acid in the presence of monochloroacetic acid, *USA Patent No.* 2 863 917, December 9, 1958.

[6] Zanaveskin, L.N., Averianov, V.A., Kunitsyn, D.G. and Bulanov, V.N., Monochloroacetic acid: chemistry, technology, perspectives, *Russian Chemical Industry, No.* 5, pp. 28–42, 2001.

[7] Zanaveskin, L.N., Bulanov, V.N., Kunitsyn, D.G., Uskach, Ya.L., Milgotin, I.M. and Varshaver, E.V., Production of monochloroacetic acid. Report 2: Processing of side dichloroacetic acid by hydrogenation in stationary catalyst layer, *Catalysis in Industry, No.* 4, pp. 29–35, 2003.

[8] Kunitsyn, D.G., Reactors with fixed-bed catalyst for dichloroacetic processing, *PhD Dissertation* (Federal Scientific Research Institute "Syntez", Moscow), 2003, 187 p.

[9] Renon, H. and Prausnitz, J. M., Local Composition in Thermodynamic Excess Functions for Liquid Mixtures, *AIChE J.*, Vol. 14, *No.* 1, pp. 135–144, 1968.

[10] Hayden, J.G. and O'Connell, J.P., A Generalized Method for Predicting Second Virial Coefficients, *Ind. Eng. Chem. Proc. Des. Dev.*, Vol. 14, *No.* 3, pp. 209–216, 1975.

In: New Aspects of Biochemical Physics
Editors: Sergei D. Varfolomeev et al., pp. 41-51
ISBN 1-60021-419-3

Chapter 6

STEADY-STATE MULTIPLICITIES IN CONTINUOUS PROCESS OF EQUILIBRIUM OPEN EVAPORATION WITH IRREVERSIBLE AND REVERSIBLE CHEMICAL REACTION

D. M. Birukov**, *E. A. Glushachenkova and Yu. A. Pisarenko
M. V.Lomonosov Moscow State Academy of Fine Chemical Technology,
pr. Vernadskogo, 86, Moscow, 119571, Russia,
Department of Chemistry and Chemical Engineering for Basic Organic Synthesis

ABSTRACT

There have been studied steady-state multiplicities in the continuous process of equilibrium open evaporation with an irreversible and reversible chemical reaction on the grounds of the software written in the *MatLab* environment. In case of an irreversible reaction there have been obtained results confirming deductions gained earlier under an analytical investigation of the similar process. It has been shown, that the type of a stable stationary state depends, besides operating parameters, on the system background. There has been got the criterion, permitting to predict an existence of several steady modes at the fixed process operating parameters on the base of information about system physicochemical properties.

Key words: Steady state multiplicities, irreversible and reversible chemical reaction, reactive distillation, phase diagram, singular points, nodes, saddles, complicated stationary state, bifurcation diagram, continuous equilibrium open evaporation, system background, residue curve map.

* Tel.: 7 (495) 936-8905, Fax: 7 (495) 434-8711; E-mail: bmi@sumail.ru

NOMENCLATURE

F	feed or product stream rate;
i and j	component numbers, i, j=1,2,3; $i \neq j$;
k_1	direct reaction rate constant;
k_2	indirect reaction rate constant;
$r(\vec{X})$	specific reaction (1) or (2) rate as a function of the composition;
R_1, R_2	reagents;
R_3	product;
t	time;
V	liquid volume inside the system;
x_i	concentration of a component i in the liquid inside the system;
x_i^F	concentration of a component i in the liquid stream, feeding the system;
$\vec{X}$	composition of the liquid inside the system;
y_i^D	concentration of a component i in the vapor stream leaving the system;
α_{i3}	relative volatility of component i to 3;
α_{j3}	relative volatility of component j to 3;
β	bifurcation parameter;
ν_i	stoichiometric coefficient, positive for products and negative for reagents;
ξ	reaction extent.

INTRODUCTION

Growing interest of modern researchers in reactive distillation (combined) processes is caused by their economic and technological advantages in comparison with their traditional analogues supposing to carry out a chemical reaction and to separate a reactionary mixture in distinct devices [1-3].

Investigations have shown that complex behaviour dynamics inhere in reactive distillation processes due to a mutual influence of distillation and chemical constituents at the expense both of concentration and energetic factors.

In a number of cases [5-8,9,10] this results in the occurrence of multiple stationary modes (multiple steady states) of operating of a device at fixed working parameters, that considerably complicates the process management.

The set of possible stationary states and their character determine limit opportunities of a process, a choice of start-up and shut-down modes and also the design task of an automatic process control system and the operational safety of a unit as a whole. Thus, there appears the problem of a choice of a strategy of achieving the most preferable stationary state.

Thus, researching of steady states multiplicities and a mechanism and conditions of its origin is an urgent task being of direct practical interest.

EXPERIMENTAL

In the present work an elementary reactive distillation process has been chosen as a research object – the continuous equilibrium open evaporation with an irreversible and reversible chemical reaction.

A choice of a simplified model is explained by the complexity of combined processes and, as a consequence by the absence of an opportunity of a qualitative analysis of their models.

Nevertheless, the common regularities revealed in such systems earlier under steady state multiplicities investigation [9,10] also inhere in more complex reactive distillation processes. For example, the computational experiment [11] has shown the presence of several stationary operating modes of the trayed reactive distillation column with the reaction zone localized on several plates.

The present work consists of several steps:

- derivation of the mathematical model of the process for the case of an irreversible and reversible reaction;
- determination of conditions of the steady state multiplicities origin;
- development of the software for steady state observing and mapping of process trajectories reflecting its changes in time;
- investigation of the evolution of system phase diagrams;
- allocation of stationary state stability areas.

Consider each of the above listed stages in more detail.

When deriving of the mathematical model of the equilibrium open evaporation with the chemical reaction

$$R_1 + R_2 \xrightarrow{k_1} R_3 \text{ or} \quad (1)$$

$$R_1 + R_2 \underset{k_2}{\overset{k_1}{\rightleftarrows}} R_3, \quad (2)$$

where R_1, R_2 – reagents; R_3 – product; k_1 and k_2 – constants of a direct and indirect reaction rate accordingly, the following assumptions have been accepted:

- components relative volatility coefficients are constant;
- mixture is zeotropic and ternary;
- liquid volume is constant in the system;
- molar density of the mixture is constant and equal to unit;
- liquid in the bottom is uniform in its physicochemical properties;
- liquid and vapor leaving the system are in equilibrium;
- reaction order is equal to unit for all the components;
- constant of irreversible reaction rate is equal to unit.

The concentration simplex, in this case, the Gibbs triangle, acts as a dynamic system phase space.

For the system of equilibrium open evaporation with the chemical reaction (1) or (2) write down the material balance for a component i for the contour I (Figure 1):

$$d\left(Vx_i\right)=\left[F\left(x_i^F-y_i^D\right)+\nu_i Vr\left(\overrightarrow{X}\right)\right]dt \tag{3}$$

where x_i – concentration of a component i $\left(i=\overline{1,3}\right)$ in the liquid with a volume V inside the system; F – rate of the feed and product streams; x_i^F – concentration of a component i in the liquid stream F, feeding the system; y_i^D – concentration of a component i in the vapor stream F leaving the system; ν_i – stoichiometric coefficient, positive for products and negative for reagents; $r\left(\overrightarrow{X}\right)$ – specific rate of the reaction (1) or (2) as a function of the liquid composition $\overrightarrow{X}$ with a volume V inside the system; t – time.

Expression (3) produces the differential equation for a component i:

$$\frac{dx_i}{dt}=\beta\left(x_i^F-y_i^D\right)+\nu_i r\left(\overrightarrow{X}\right) \tag{4}$$

where the parameter $\beta = F/V$.

So the mixture is ternary and at any point of time the term

$$d\left(x_1+x_2+x_3\right)=0 \tag{5}$$

holds true, it's enough to have two equations like (4) to describe the system state at any moment of time.

Supposing that the reactionary mixture is ideal and component relative volatilities are constant, get the correlation, connecting the concentration of a component i in the liquid phase x_i with its equilibrium concentration of a component i in the vapor phase y_i^D:

$$y_i^D=\frac{\alpha_{i3}x_i}{\left(\alpha_{i3}-1\right)x_i+\left(\alpha_{j3}-1\right)x_j+1} \tag{6}$$

where i and j – component numbers, i, j=1,2; $i \neq j$; α_{i3} and α_{j3} – relative volatilities of component i to 3(j to 3) respectively.

Reasoning from the above accepted assumptions, $r\left(\overrightarrow{X}\right)$ for the irreversible reaction (1) can be expressed by the equation:

$$r\left(\overrightarrow{X}\right)=x_1x_2 \tag{7}$$

and $r(\overrightarrow{X})$ for the reversible reaction (2) – by the following one:

$$r(\overrightarrow{X}) = k_2 \left(K_P x_1 x_2 - x_3^2\right) \tag{8}$$

where $x_3 = 1 - x_1 - x_2$ and the equilibrium constant $K_P = {k_1}/{k_2}$.

Then, taking into consideration expressions (4) – (7) for the irreversible reaction (1) it's possible to write down the following system of differential equations:

$$\begin{cases} \dfrac{dx_1}{dt} = \beta\left(x_1^F - \dfrac{\alpha_{13}x_1}{(\alpha_{13}-1)x_1 + (\alpha_{23}-1)x_2 + 1}\right) - x_1x_2 \\ \dfrac{dx_2}{dt} = \beta\left(x_2^F - \dfrac{\alpha_{23}x_2}{(\alpha_{13}-1)x_1 + (\alpha_{23}-1)x_2 + 1}\right) - x_1x_2 \end{cases} \tag{9}$$

where $\beta = {F}/{V}$,

and in view of expressions (4) – (6) and (8) for the reversible reaction (2) – the system

$$\begin{cases} \dfrac{dx_1}{d\tau} = \beta'\left(x_1^F - \dfrac{\alpha_{13}x_1}{(\alpha_{13}-1)x_1 + (\alpha_{23}-1)x_2 + 1}\right) - \left[K_P x_1 x_2 - (1 - x_1 - x_2)^2\right] \\ \dfrac{dx_2}{d\tau} = \beta'\left(x_2^F - \dfrac{\alpha_{23}x_2}{(\alpha_{13}-1)x_1 + (\alpha_{23}-1)x_2 + 1}\right) - \left[K_P x_1 x_2 - (1 - x_1 - x_2)^2\right] \end{cases} \tag{10}$$

where $\tau = k_2 t$ and $\beta' = {F}/{k_2 V}$.

Thus, the systems of equations (9) and (10) describe changing of the process state with time. Stationary states are characterized by constancy of process parameters. Hence, on putting to zero the right part of (4), we get

$$F\left(x_i^F - y_i^D\right) = \nu_i V r(\overrightarrow{X}) = \nu_i \xi' = \nu_i \frac{\xi'}{F} F = \nu_i \xi F \tag{11}$$

where ξ – reaction extent, and so components 1 and 2 are reagents, then $\nu_1 = \nu_2 = -1$.

Relying on (11) for a steady state in case of the irreversible reaction it can be written, that

$$r(\overrightarrow{X}) - \beta\xi = 0 \tag{12}$$

or, in case of the reversible reaction

$$\left(r\left(\overrightarrow{X}\right) \Big/ k_2 \right) - \beta'\xi = 0 \tag{13}$$

Now, using relations (6) and (11), express x_1 and x_2 as functions of ξ:

$$x_1 = x_1\left(\xi\right) = \frac{\alpha_{23}\left(x_1^F - \xi\right)}{\alpha_{13}\alpha_{23} - \alpha_{13}\left(x_2^F - \xi\right)\left(\alpha_{23} - 1\right) - \alpha_{23}\left(x_1^F - \xi\right)\left(\alpha_{13} - 1\right)} \tag{14}$$

$$x_2 = x_2\left(\xi\right) = \frac{\alpha_{13}\left(x_2^F - \xi\right)}{\alpha_{13}\alpha_{23} - \alpha_{13}\left(x_2^F - \xi\right)\left(\alpha_{23} - 1\right) - \alpha_{23}\left(x_1^F - \xi\right)\left(\alpha_{13} - 1\right)} \tag{15}$$

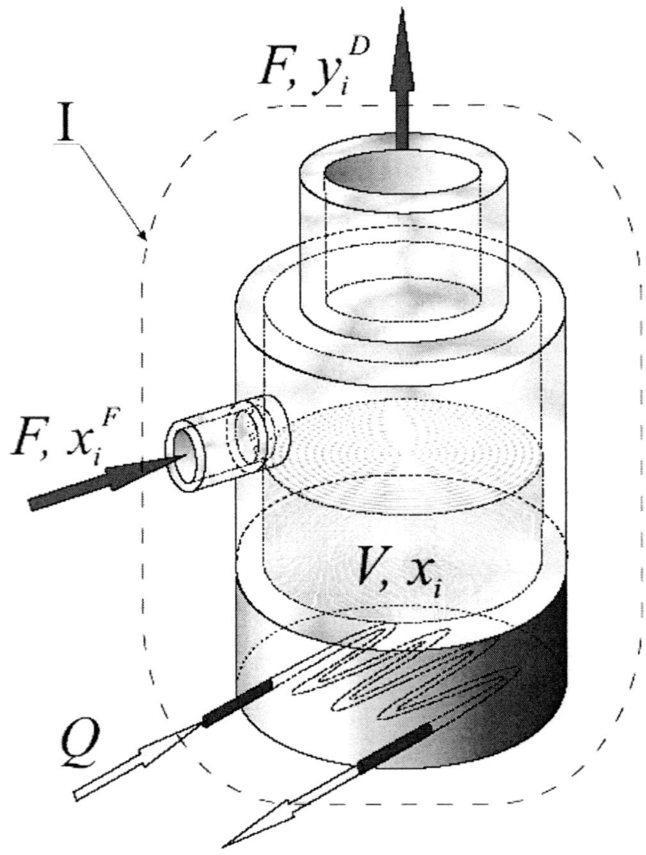

Figure. 1. Scheme of material streams of the continuous equilibrium open evaporation with a chemical reaction: F – rate of feed and product streams; xiF – concentration of a component i $\left(i = \overline{1,3}\right)$ in the liquid stream F, feeding the system; yiD – concentration of a component i in the vapor stream F leaving the system; xi – concentration of a component i in the internal system liquid of a volume V, Q – heat supply.

Next, on the grounds of equations (15) and (16) obtained and equality (7) for the irreversible reaction, write down r as a function of ξ and put the expression gained for $r(\xi)$ in (13).

As a result we get:

$$\underbrace{\frac{\alpha_{13}\alpha_{23}(x_1^F-\xi)(x_2^F-\xi)}{\left[\alpha_{13}\alpha_{23}-\alpha_{13}(\alpha_{23}-1)(x_2^F-\xi)-\alpha_{23}(\alpha_{13}-1)(x_1^F-\xi)\right]^2}}_{r(\xi)}-\beta\xi=0 \tag{16}$$

In a similar manner, for the reversible reaction, using equations (15), (16) and (8), and taking into consideration that in the feed mixture $x_1^F+x_2^F=1$, express r as a function of ξ and put the equation gained for $r(\xi)$ in (14) instead of $r\left(\overrightarrow{X}\right)$. At that we gain the following expression:

$$\underbrace{\frac{\alpha_{13}\alpha_{23}\left[K_P(x_1^F-\xi)(x_2^F-\xi)-4\alpha_{13}\alpha_{23}\xi^2\right]}{[\alpha_{13}\alpha_{23}-\alpha_{13}(\alpha_{23}-1)(x_2^F-\xi)-\alpha_{23}(\alpha_{13}-1)(x_1^F-\xi)]^2}}_{r(\xi)/k_2}-\beta'\xi=0 \tag{17}$$

So the problem of defining possible process steady modes comes to determining a quantity of roots ξ of an equation like this:

$$\Phi(\xi)\equiv r(\xi)-\beta\xi=0 \tag{18}$$

for the irreversible reaction, or, for the reversible reaction, of an equation like this:

$$\Phi(\xi)\equiv\left(r(\xi)/k_2\right)-\beta'\xi=0 \tag{19}$$

The latter correspond to crossing points of the curve $r(\xi)$ and the line $\beta\xi$ in case of the irreversible reaction or the curve $r(\xi)/k_2$ and the line $\beta'\xi$ in case of the reversible reaction, presenting, accordingly, the rate of substance production inside the system and the rate of its removing to the environment. On putting obtained values of ξ in (15) and (16), it's possible to calculate coordinates of singular points.

To determine steady states and map process trajectories, in this work the software written in the environment of *MatLab* [12] has been used.

RESULTS

It's convenient to summarize the results of the investigation of continuous equilibrium open evaporation as the diagram set, permitting to follow the residue curve map evolution of a corresponding dynamic system when varying any bifurcation parameter. As an example, in Figure 2 there has been illustrated the evolution of residue curve maps of the system in case of the irreversible reaction (1) at $\alpha_{13} = 0.002$; $\alpha_{23} = 0.200$; $x_1^F = 0.30$; $x_2^F = 0.70$ (in mole fractions) and changing β from 0.21 to 0.66.

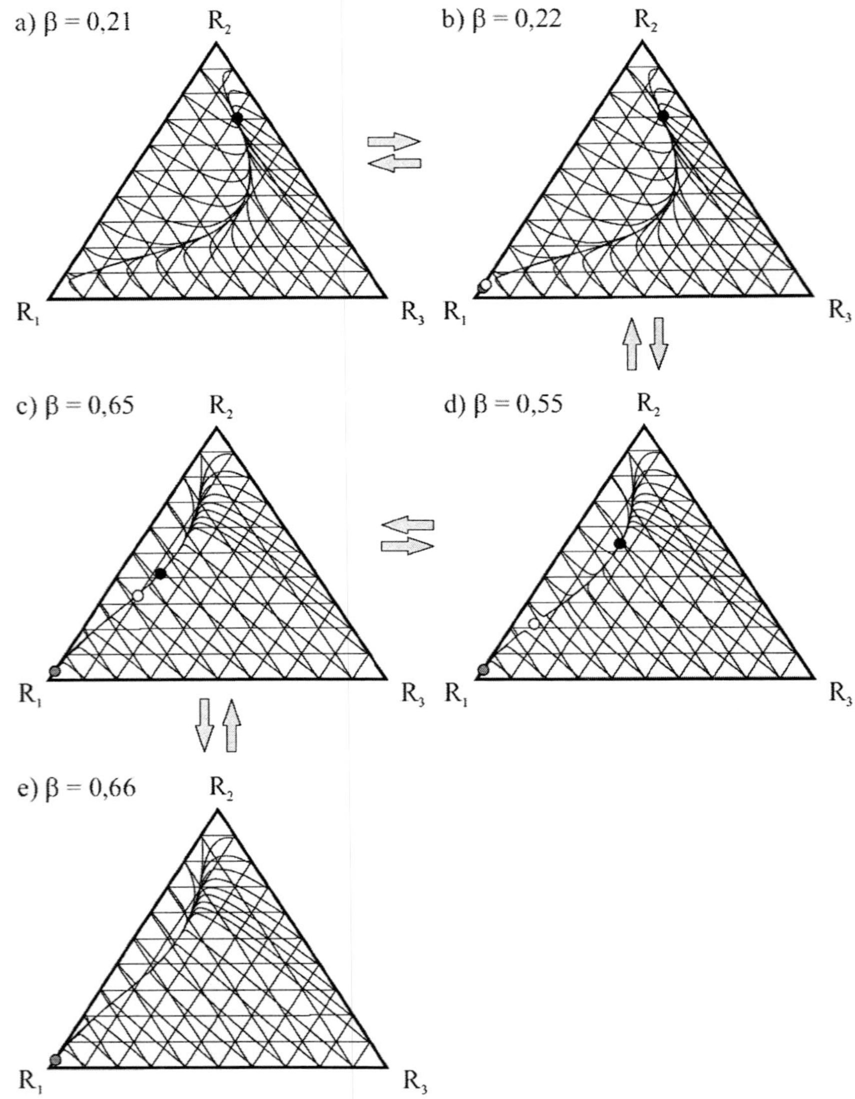

Figure 2. Residue curve map evolution of continuous equilibrium open evaporation with irreversible reaction $R_1 + R_2 \rightarrow R_3$ at α13 = 0.002; α23 = 0.200; x1F = 0.30 and x2F = 0,70 (in mole fraction)

Consider the results obtained in more details.

The diagram for $\beta = 0.21$ presented in Figure 2a is characterized by the single steady state and the singular point in the concentration simplex is a topological node. While increasing of a value β to 0.22, the system becomes unstable, as a result the complicated topological structure of the saddle node is generated (Figure 2b). When β mounts to 0.65, tree singular points have been formed (Figure 2c), two of them are topological nodes, and one located between them is a saddle, i. e. this case agrees with the set of three stationary states. Under subsequent rising of the operating parameter β the structure is becoming unsteady again (Figure 2d), and at a value β of 0.66 in the diagram the single stationary state remains, corresponding to a topological node (Figure 2e). Thus, varying the operating parameter β has an influence on the number and type of steady states at the fixed feed composition.

The mutual arrangement of $r(\xi)$ and $\beta\xi$ (or $r(\xi)/k_2$ and $\beta'\xi$, expressions (19) and (20)) is determined by physicochemical properties of the mixture (α_{13}, α_{23}), and also by values of the so-called bifurcation parameters: x_i^F and β for the irreversible and x_i^F, β', K_P for the reversible reaction. In this connection it's of interest to investigate the parametritic spaces of (x_i^F; β) and (x_i^F, β', K_P) in order to reveal the regions of existing of multiple steady states.

As an example, in Figure 3 the bifurcation diagram for the continuous equilibrium open evaporation with the irreversible reaction (1) is depicted. In case of the reversible reaction (2) the bifurcation diagram is presented by two surfaces dividing the parametritic space (x_i^F, β', K_P) into three regions with a different quantity of steady states.

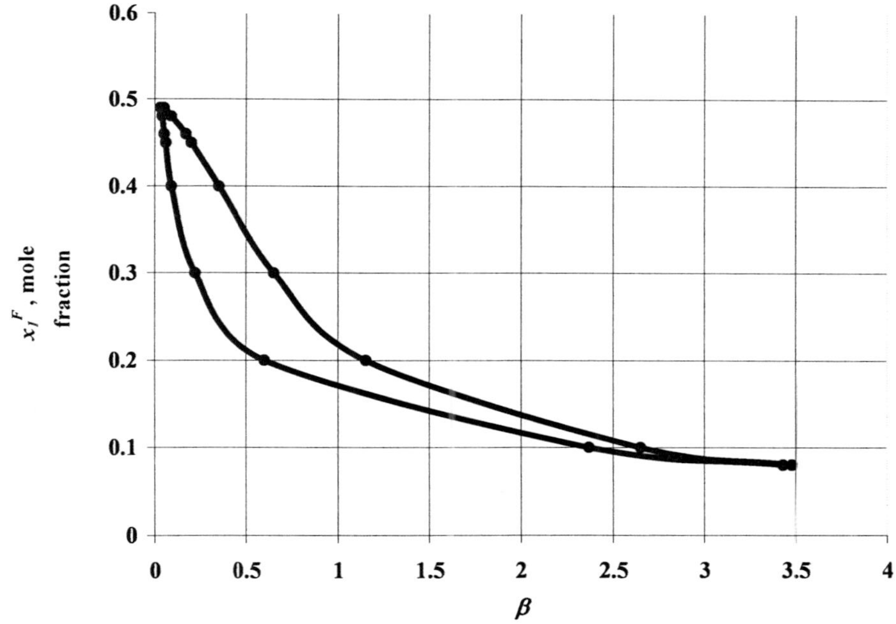

Figure. 3. Bifurcation diagram of continuous equilibrium open evaporation with irreversible reaction $R_1 + R_2 \rightarrow R_3$

The results obtained (Figure 3) have demonstrated, that at the certain values of operating parameters (x_i^F and β for the irreversible reaction and x_i^F, β', K_P for reversible one) in the parametritic space two curves stand out (surfaces in case of the reversible reaction), corresponding to the upper and lower values of the bifurcation parameter, with those, in its turn, the complicated singular points comply (Figure 2) in the phase diagrams. Three stationary states correspond to the bifurcation diagram internal region included between curves (surfaces). The single steady state agrees with the points, not belonging to the edges or the internal region of the bifurcation diagrams.

CONCLUSIONS

Hence, on the grounds of the software written in the *MatLab* environment there has been investigated a phenomenon of steady-state multiplicities in the continuous equilibrium open evaporation with the irreversible and reversible reaction. In case of the irreversible reaction there have been obtained results confirming deductions gained earlier under an analytical investigation of the similar process [9,10]. It has been established, that in the presence of multiple steady states the character of a stable stationary state depends, besides operating parameters, on the system background. There has been shown that coming of operating process parameters over some critical values leads to a sharp change of the system behaviour owing to the realignment of its phase diagram structure, at that the number and type of singular points are modified. There has been got the criterion, permitting to predict the possibility of an existence of several steady modes at the fixed process operating parameters on the base of information about physicochemical system properties.

Financial support for this research is provided by the Russian Foundation for Basic Research, the project *No* 05-03-32899.

REFERENCES

[1] Pisarenko, Yu.A., Cardona, K.A. and Serafimov, L.A., *Reactsionno-rectifikatsionnie protsessi: dostijeniya v oblasti issledovaniya i practicheskogo ispol'zovaniya* (Reactive distillation processes: advances in research and practical application) (Luch Moscow), 2001, 268 p. *(in Russian)*

[2] Pisarenko, Yu.A., Glushachenkova, E.A., Shuvalov, A.S., Timoshenko, A.V., Borutskiy, P.N., Podkletnova, N.M. and Dorogochinskaya, V.A., Method for producing of high-octane gasoline additive, *RU Certificate (Patent) No.* 2 248 342, December 29, 2003.

[3] Pisarenko, Yu.A. and Cardona, K.A., Method for producing of alkyl acetates, *RU Certificate (Patent) No.* 2 204 549, June 26, 2001.

[4] Sundmacher, K. and Kienle, A., *Reactive Distilation. Status and Future Directions* (Wiley VCH Verlag GmbHB Co. KGaA., Weinheim), 2003, 287 p.

[5] Hauan, S., Hertzberg, T. and Lien, K.M., Why methyl tret-butyl ether production by reactive distillation may yield multiple solutions, *Ind. Eng. Chem. Res.*, Vol. 34, *No* 3, pp. 987–991, 1995.

[6] Ciric, A.R. and Miao, P., Steady state multiplicities in an ethylene glycol reactive distillation column, *Ind. Eng. Chem. Res.*, Vol. 33, *No* 11, pp. 2738–2748, 1994.

[7] Mohl, K.D., Kienle, A. and Gilles, E.D. Multiple steady states in a reactive distillation column for the production of the fuel ether TAME. I. Theoretical analysis, *Chem. Eng. Technol.*, Vol. 21, *No* 2, pp. 133–136, 1998.

[8] Rapmund, P., Sundmacher, K. and Hoffmann, U., Multiple steady states in a reactive distillation column for the production of the fuel ether TAME. II. Experimental validation, *Chem. Eng. Technol.*, Vol. 21, *No* 2, pp. 136-139, 1998.

[9] Pisarenko, Yu.A., Epifanova, O.A. and Serafimov, L.A., Investigation of the process dynamics of the continuous evaporation with a chemical reaction, *Theor. Found. Chem. Eng.*, Vol. 22, *No* 6, pp. 723–728, 1988.

[10] Pisarenko, Yu.A. and Serafimov, L.A., Steady state multiplicities in the process of the equilibrium continuous evaporation with a chemical reaction, *Theor. Found. Chem. Eng.*, Vol. 26, *No* 6, pp. 888–892, 1992.

[11] Pisarenko, Yu.A., Anokhina, E.A. and Serafimov, L.A., Searching of the steady state multiplicity of countercurrent reactive distillation processes, *Theor. Found. Chem. Eng.*, Vol. 27, *No* 6, pp. 586–590, 1993.

[12] Anufriev, I.E., Samouchitel’ po MatLab 5.3/6.x. (Manual for self-tuition in MatLab 5.3/6.x.) (BVH-Peterburg SPb), 2004, 736 p.

In: New Aspects of Biochemical Physics
Editors: Sergei D. Varfolomeev et al., pp. 53-58
ISBN 1-60021-419-3

Chapter 7

CHOICE OF OPTIMAL DISTILLATION-BASED SEQUENCE FOR SEPARATION OF LIGHT GASOLINE C_{4-6+} FRACTION

A. S. Kinyakin*, E. A. Glushachenkova and Yu. A. Pisarenko
M. V.Lomonosov Moscow State Academy of Fine Chemical Technology,
pr. Vernadskogo, 86, Moscow, 119571, Russia,
Department of Chemistry and Chemical Engineering for Basic Organic Synthesis

ABSTRACT

There have been compared nine distillation-based sequences for the light gasoline C_{4-6+} fraction separation, including conventional and complex columns on basis of the criterion of the total heating steam and cooling water costs minimum. It has been shown that the direct distillation column arrangement is the optimal one for the considered feed composition and product fraction purities of 90.0–98.0 wt. %. It has been observed that the direct sequence remains optimal when varying the feed composition.

Keywords: Light gasoline C_{4-6+} fraction isomerization, contemporary industrial processes of n-pentane and hexanes isomerization, structural and parametritic optimization of distillation unit, optimal distillation column arrangement, direct (indirect) distillation column sequence, column with side stream draw-off, column with side stripper (rectifier), column with (pre)factionator and side stream draw-off, total heating steam and cooling water costs.

* Tel.: 7 (495) 936-8905, Fax: 7 (495) 434-8711. E-mail: Kinyakin@mail.ru

INTRODUCTION

The problem of environment pollution by exhaust gases has made for bringing in new more rigid requirements to commercial and ecological gasoline specifications [1], in particular implying elimination of aromatics and olefins concentrations in fuel.

This fact restricts cracking and pyrolysis application for the high-octane gasoline additives production and results in growing of the light gasoline C_{4-6+} (pentane-hexane C_{5-6}) fraction isomerization role in increasing of the branched isomer concentration in it [2].

The discrepancy between conditions, under which the isomerization of *n*-pentane and hexanes passes, causes the need of separating the raw C_{4-6+} (C_{5-6}) fractions into individual cuts, containing mainly a single component, that is carried out in contemporary industrial processes [3, 4] by the methods of adsorption on molecular sieves or distillation. The latter is characterized by extensive process flowsheets and high operating costs coming up to 70% of total running costs of a plant and being among the factors which form the prime production cost.

One way to reduce gasoline production operating costs (involving heating steam and cooling water costs) and to raise their competitive capacity is structural and parametritic optimization of C_{4-6+} or C_{5-6} fraction distillation units. The present work is devoted to solving of this problem.

EXPERIMENTAL

As an object of optimization there has been considered the distillation unit involved in the process flowsheet of reactive distillation izomerization of the light gasoline C_{4-6+} fraction. In that case, as in well-known industrial processes [3, 4], the separate isomerization of pentane and hexane fractions is also provided, but there are made less rigid demands of C_4 and C_7 impurities content in them. So the sufficient condition is to separate the raw C_{4-6+} gasoline fraction into only three fractions: butane-pentane (C_{4-5}), hexane (C_6) and hexane-heptane (C_{6+}) cuts.

In this work there has been fulfilled the comparison of nine distillation column arrangements (Figure 1) by the value of total heating steam and cooling water costs at a feed charge of 360 thousand tons/year. The raw is the light gasoline C_{4-6+} fraction (*the base mixture*) with the following composition (wt. %): *iso*-butane – 0.10; *n*-butane – 2.83; *iso*-pentane – 17.26; *n*-pentane – 24.82; cyclopentane – 2.85; *2*-methylpentane – 11.01; *3*-methylpentane – 6.48; *n*-hexane – 14.84; methylcyclopentane – 8.39; cyclohexane – 5.01; *n*-heptane – 3.45; *1*-*cis*-*2*-dimetilcyclopentane – 2.96.

Product fraction (C_{4-5}, C_6 and C_{6+}) purities are selected on the basis of technical specifications and amount to 90.0, 95.0 or 98.0 wt. %. The feed is sent into the distillation unit in the form of boiling liquid. Heating steam at a pressure of 0.18–0.42 MPa is used as a heat-transfer agent in column reboilers, and reused water at an inlet temperature of 25.0° C – as a refrigerant in column condensers. At that a pressure of heating steam and an outlet temperature of reused water are selected in such a way as to provide the temperature difference of 10.0° C between the condenser (reboiler) outlet streams.

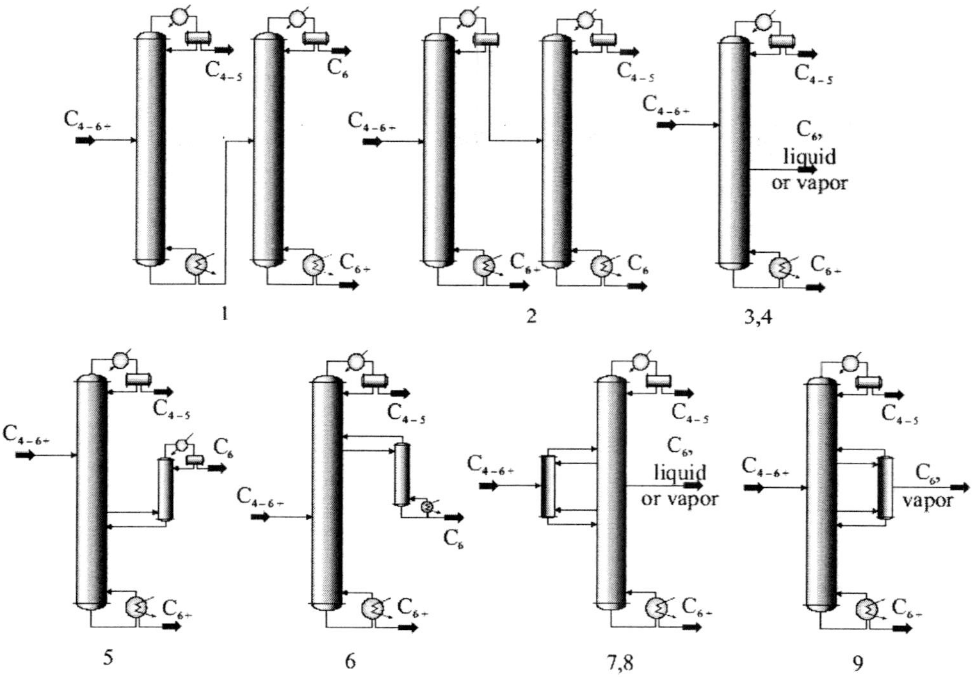

Figure. 1. Distillation column arrangements for light gasoline C_{4-6+} fraction separation: 1 – direct sequence; 2 – indirect sequence: 3,4- column with liquid or vapor side stream draw-off accordingly; 5 - column with side rectifier; 6 - column with side stripper; 7,8 – column with prefractionator and liquid (vapor) side stream draw-off accordingly; 9 – column with fractionator having vapor side stream draw-off; C_{4-6+} – feed stream; product fractions: C_{4-5} (butanes, n-pentane, iso-pentane, cyclopentane), C_6 (n-hexane, its isomers and methylcyclopentane), C_{6+} (cyclohexane and heavier)

The estimation of sequences 1–9 has been carried out with use of the *Inside-Out* theoretical tray-based algorithm designed for the distillation column simulation and involved in the proprietary software of *PRO/II.* Conversion from theoretical trays into actual ones is effected by means of a tray efficiency factor. In all the columns there have been accepted valve trays with an efficiency factor of 0.65. The vapor-liquid equilibrium has been simulated by the Soave-Redlich-Kwong equation of state [5].

Optimized parameters of conventional columns (sequences 1 and 2) are: the number of theoretical and actual trays, an operating pressure and the feed tray location. For the columns with the side stream draw-off (sequences 3 and 4) the draw-off tray location is added to the above-listed parameters. In the sequences 5–9 there have been extra optimized the ratio of the main column tray number to the side section tray number, the feed and draw-off tray location and the points of inputting and withdrawing of streams connecting the main column and the side section.

RESULTS

The results of calculation of the total heating steam and cooling water costs of the sequences 1–9 at optimal constructional and operating parameters and the identical summary actual tray number of 111 are depicted in Figure 2.

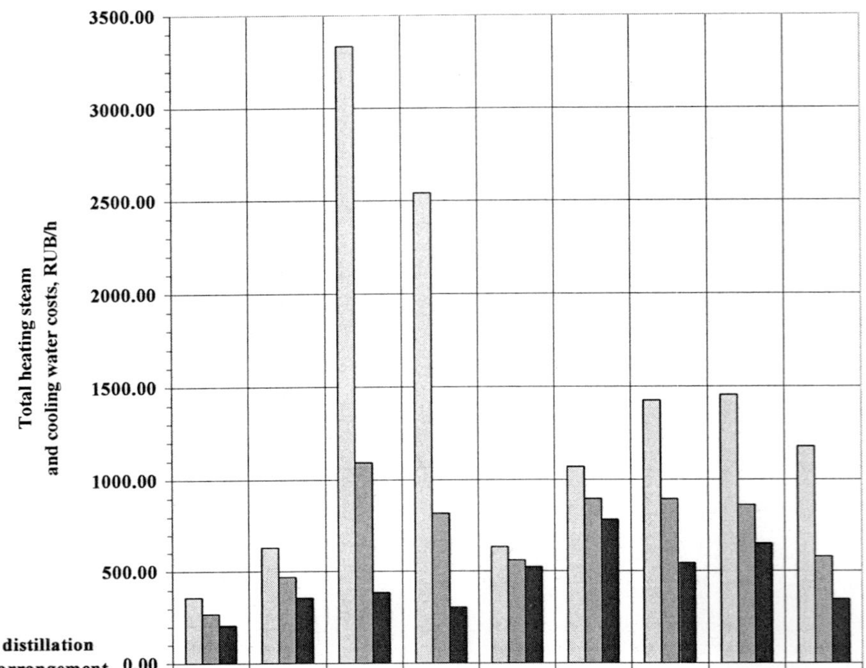

No. of distillation column arrangement	1	2	3	4	5	6	7	8	9
□ 98.0%, RUB/h	359.25	629.00	3333.50	2540.90	631.77	1069.49	1425.00	1454.26	1177.18
▨ 95.0%, RUB/h	270.24	468.86	1092.66	819.36	558.66	895.89	894.79	864.03	577.46
■ 90.0%, RUB/h	207.17	358.54	386.28	306.53	522.51	785.55	541.95	648.93	348.90

Figure. 2. Total heating steam and cooling water costs for product fraction purities of 90.0, 95.0 and 98.0 wt. % and base mixture: 1 – direct sequence; 2 – indirect sequence: 3,4- column with liquid or vapor side stream draw-off accordingly; 5 - column with side rectifier; 6 - column with side stripper; 7,8 – column with prefractionator and liquid (vapor) side stream draw-off accordingly; 9 – column with fractionator having vapor side stream draw-off

The direct sequence 1 corresponds to the least utility costs. Product fraction purity reduction from 98.0 to 90.0 % has led to total heating steam and cooling water costs decrease that is the most essential (up to 88.0 %) for the arrangements 3 and 4. At product fraction purities of 95.0 and 98.0 % it's possible to arrange the sequences considered in the order of increasing of total utility costs in the following way: 1, 2, 5, 6, 9, 7, 8, 4, 3, and at purities of 90.0 % – in the way: 1, 4, 9, 2, 3, 5, 7, 8, 6, i. e. the sequence 4 has moved to the second position. Nevertheless, when varying product fraction purities within the limits of 90.0–98.0 wt. %, the sequence 1 remains optimal.

In the working environment it's difficult enough to escape fluctuations of the feed C_{4-6+} fraction composition. This is the result of a number of factors such as changing of raw sources or operating modes of previous process stages. So in the present work structural stability of the optimal sequence has been investigated at the componentwise variation of the feed composition within the limits of ±5.0 wt. % (of the appropriate component concentration in *the base mixture* at the constant concentration ratio of the others) and product fraction purities of 90.0, 95.0 or 98.0 %. The above-pointed variation range of the raw composition has been chosen taking into account light gasoline C_{4-6+} fraction technical specifications. For varying of the C_{4-5} concentration in the raw at product fraction purities of 98.0 % the results are presented in Figure 3.

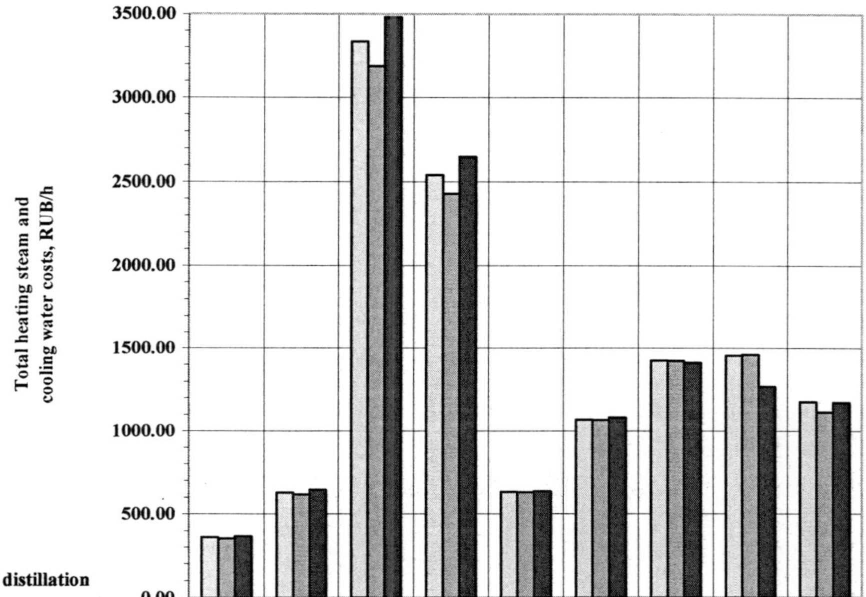

No. of distillation column arrangement	1	2	3	4	5	6	7	8	9
□ Composition No 1, RUB/h	359.25	629.00	3333.50	2540.90	631.77	1069.49	1425.00	1454.26	1177.18
▨ Composition No 2, RUB/h	353.31	615.44	3187.41	2429.07	629.89	1065.742	1423.45	1461.76	1114.75
■ Composition No 3, RUB/h	366.06	644.78	3482.17	2648.52	635.493	1083.10	1411.31	1267.26	1171.87

Figure 3. Total heating steam and cooling water costs for product fraction purities of 98.0 % and varying of C_{4-5} concentration in raw C_{4-6+} fraction: composition No. 1 – base mixture; composition No. 2 – increasing by 5.0 % of C_{4-5} concentration in comparison with base mixture; composition No. 3 – – decreasing by 5.0 % of C_{4-5} concentration in comparison with base mixture; composition No. 2 and No.3 –concentration ratio of the other components is the same as in base mixture

The similar data have been obtained for product fraction purities of 95.0 and 90.0 wt. % and varying of the C_6 and C_{6+} concentration in the raw.

The calculation investigation has demonstrated that the sequence 1 remains optimal when varying the raw composition regardless of product fraction purities.

Conclusions

Consequently, the direct sequence 1 is characterized by the least total heating steam and cooling water costs. It remains optimal when varying the raw composition within the limits established by technical specifications at product fraction purities within the bounds of 90.0–

98.0 %. Decreasing of product fraction purities form 98.0 to 90.0 % leads to reduction of the complex distillation column sequences costs and their approaching to the level obtained for the direct sequence.

Financial support for this research is provided by the Russian Foundation for Basic Research, the project *No.* 05-03-32899.

References

[1] Danilov, A.M., Na tret'ei mejdunarodnoi konferentsii "Novie topliva s prisadkami" (At the third International conference "New fuel with additives"), *Neftepererabotka i neftekhimiya*, No. 10, pp. 33–37, 2004. (in Russian).

[2] Somov, B.E., Sadchikov, I.A. and Shershun, V.G., *Strategicheskie prioriteti rossiiskikh neftepererebativauschikh predpriyatii* (Strategic priorities of Russian petroleum refineries) (TsNIITEneftekhim Moscow), 2002, 292 p. (in Russian).

[3] Domerg, B., Vatripon, L., Joly, J.-F. and Matteus, R., Dal'neichee razvitie tekhnologii izomerizatsii parafinov (Further development of isomerization technology of paraffins), *Neftepererabotka i neftekhimiya*, No 4, pp. 15–27, 2001. (in Russian).

[4] Bursian, N.R., *Tekhnologiya izomerizatsii parafinovikh uglevodorodov* (Isomerization technology of paraffin hydrocarbons) (Khimiya Leningrad), 1985, 192 p. (in Russian)

[5] Soave, G. Equilibrium constants from a modified Redlich-Kwong equation of state, *Chem. Eng. Sci.*, Vol. 27, No 6, pp. 1197–1203, 1972.

In: New Aspects of Biochemical Physics
Editors: Sergei D. Varfolomeev et al., pp. 59-62

ISBN 1-60021-419-3

Chapter 8

ELASTIC CAPSULE FILLED WITH MAGNETIC FLUID IN AN ALTERNATIVE MAGNETIC FIELD

F. S. Bayburtskiy [1], V. A. Naletova [2], V. A. Turkov [2], B. A. Mayzelis[4], E. I. Minsker [4], G. V. Stepanov [4], K. Zimmermann [5] and I. Zeidis [5]

[1] N. M. Emanuel Institute of Biochemical Physics RAS, Russian Federation, Moscow, 119991, Kosygina street, 4. E – mail: Chembio@sky.chph.ras.ru, Bayburt@mail.ru

[2] Institute of Mechanics, Lomonosov Moscow State University, Russian Federation, Moscow, 119192, Michurinsky prospectus, 1.

[3] Research Institute of Chemistry and Technology of Hetero-organic Compounds, Russian Federation, Moscow, 111123, Shosse Entuziastov, 38. E – mail: Eoe@eoe.incotrade.ru, Gstepanov@mail.ru

[4] Faculty of Mechanical Engineering, Technische Universität Ilmenau, Federal Republic Germany Germany, Ilmenau, 98684, PF 10 05 65.

ABSTRACT

The motion of a magnetizable elastic body in an alternate magnetic field is experimentally studied. The body of a cylindrical from is located in a cylindrical channel. Traveling nonuniform magnetic field is generated by an electromagnetic system. Agreement of experimental data with theoretical results has been achieved.

Key words: Magnetic fluid, magnetic field, locomotive device, deformation of magnetizable materials.

INTRODUCTION

The realization of locomotion systems using deformation of magnetizable materials (a magnetic fluid in a elastic capsule or amagnetizable polymer) in an applied magnetic field is a new interesting problem. In [1 – 3], the theory of a flow of layers of magnetizable fluids in a traveling magnetic field is discussed. It is shown, that the traveling magnetic field can create the rate of flow in the fluid layers. In [4], the theory of the behavior of a locomotion system using periodic deformation of a magnetizable polymer, when an alternate magnetic field operates, is discussed. The average velocity of such locomotion systems is proportional to the difference of the friction coefficients between the system and the substrate, which depends on directions of motion. In [5] the motion of a chain of spherical elastic balls filled with magnetic fluid in a cylindrical channel is experimentally studied. Such a chain moves in a magnetic field created by two permanent magnets, which are moved along the channel. In [6] the deformation and the motion of a cylindrical body (a body made by a magnetizable polymer) in an alternate magnetic field are experimentally studied. The cylindrical body, which is located in a cylindrical channel, is considered. It is found that there is an undulation of the body in a periodic magnetic field of special structure and the body moves along the channel. The theoretical estimation of the deformation of the body in an applied magnetic field and the body velocity is done [6]. In experiments the body velocity reaches 6 cm / sec [6]. The initiator of the motion is an alternate magnetic field, which forms to exterior sources (a electromagnetic system [6]). Such device has some characteristics, which allow to use this device in medicine and biology. For example, it does not contain solid details contacting with a surrounding medium.

In the present paper a motion of an elastic cylindrical capsule filled with a magnetic fluid in an alternate magnetic field is studied.

EXPERIMENTAL RESULTS

In our experiments an elastic cylindrical capsule filled with a magnetic fluid is located inside a cylindrical channel. The channel and the capsule diameters are 11 mm and 4.5 mm. The length of the capsule filled with a magnetic fluid is 75 mm.

An electromagnetic system contains coils which are along the channel. The axes of the coils are in a horizontal plane, L is the distance between the adjacent axes ($L=25$ mm). The coils are placed at the left and right sides of the channel. A magnetic field is created by three coils simultaneously. The axis of the middle coil is the symmetry axis of the magnetic field. These three adjacent coils are turned on at the same moment. Periodically the left coil is turned off and the next coil (fourth) is turned on. Thus the magnetic field travels to right along the channel. The maximum value of the magnetic field on the channel axis is about 300 Oe. The parameter n is the number of the coil switches per a second. We call n as a frequency. In our experiments the frequency changes from 2 s^{-1} to 100 s^{-1}, here- with the capsule velocity changes from 0.05 cm / s to 2.5 cm / s.

It is found that there is an undulation of the capsule in a periodic travelling magnetic field of special structure and the capsule moves along the channel. The phases of motion and the deformation of the capsule are shown in figure 1.

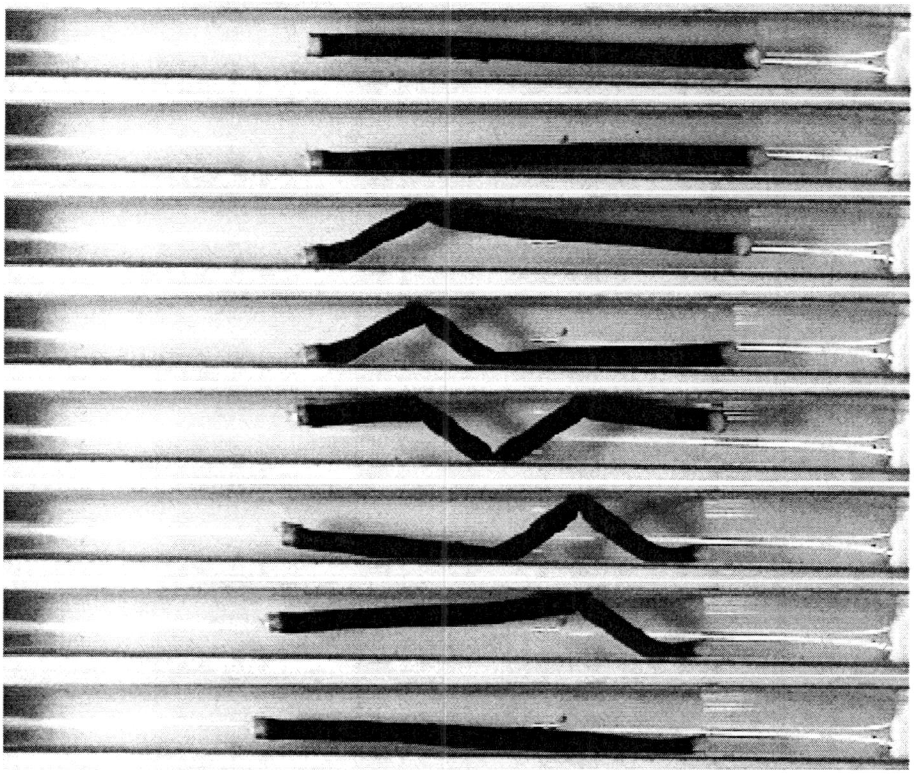

Figure 1. The form of the capsule at different moments duringa cycle of a deformation in the traveling magnetic field.

CONCLUSION

It is shown experimentally that in a specially structured periodic traveling magnetic field the elastic cylindrical capsule filled with a magnetic fluid moves along the channel. Direction of the capsule motion is opposite to the direction of the traveling magnetic field. The maximal obtained capsule velocity is 2.5 cm / s. For the frequency $n < 100\ s^{-1}$ the theoretical estimations of the velocity of the capsule agree with the experiments. The optimum geometrical sizes of the capsule and the channel are discussed.

ACKNOWLEDGMENTS

This work is supported by Deutsche Forschungsgemeinschaft (DFG, ZI 540/7-1) and the Russian Foundation for Basic Research (project 05-01-04001).

REFERENCES

[1] K. Zimmermann, I. Zeidis, V. A. Naletova, V. A. Turkov. Waves on the surface of a magnetic fluid layer in a traveling magnetic field. // *J. Magn. Magn. Mater.* 268, (2004) 227 – 231.

[2] K. Zimmermann, I. Zeidis, V. A. Naletova, V. A. Turkov. Travelling waves on a free surface of a magnetic fluid layer. // *J. Magn. Magn. Mater.* 272 – 276 , (2004) 2343 – 234.

[3] K. Zimmermann, I. Zeidis, V. A. Naletova, V. A. Turkov, V. E. Bachurin. Locomotion based on a two-layers flow of magnetizable nanosuspensions, In: Joint European Magnetic Symposia (5 – 10 September 2004, Dresden, Germany), (2004) 134.

[4] K. Zimmermann, I. Zeidis, V. A. Naletova, V. A. Turkov. Modelling of worm-like motion systems with a magneto-elastic elements. // *Phys. Stat. Solid.* 1 (12), (2004) 3706 – 3709.

[5] N. Saga, T. Nakamura. Elucidation of propulsive force of microrobot using magnetic fluid. // *J. Appl. Phys.* 91 (10), (2002) 7003 – 7005.

[6] K. Zimmermann, I. Zeidis, V. A. Naletova, G. V. Stepanov, V. A. Turkov. M.V. Lukashevich. Undulation of a magnetizable elastic body in a magnetic field. In: Book of Abstracts, Moscow International Symposium on Magnetism, June 25 – 30, 2005.

In: New Aspects of Biochemical Physics
Editors: Sergei D. Varfolomeev et al., pp. 63-67
ISBN 1-60021-419-3

Chapter 9

INFLUENCE OF PRESENCE OF FOREIGN SURFACES ON STABILITY OF MAGNETIC FLUIDS

F. S. Bayburtskiy, A. I. Vilenskiy and A. D. Korenev
N. M. Emanuel Institute of Biochemical Physics of RAS, Russian Federation, Moscow, 119991, Kosygina street, 4. E – mail: Chembio@sky.chph.ras.ru, Bayburt@mail.ru

ABSTRACT

The results of the interaction of highly dispersed magnetic particles with the surface of a foreign solid phase in different liquid media are given. On the basis of an analysis of experimental kinetic data, a phenomenological theory of the deposition process of magnetic particles on a foreign surface is developed. The dependence's of the influence of the nature and composition of liquid media on the interaction of containing solid phases in the system under investigation are established. A stability criterion for magnetic particles during interaction with a foreign surface under different conditions is proposed.

Keywords: magnetic fluids, colloidal systems, heterocoagulation, stabilization of particles.

INTRODUCTION

Magnetic fluids, widely used as the main component in many modern technical devices, must process high stability during phase separation under static and dynamic conditions [1]. Real magnetic fluids, highly dispersed heterogeneous systems, have a certain stability level with respect to aggregation of magnetic particles, defined by the conditions of their preparation and stabilization. Change of the physical-chemical characteristics of magnetic fluids with time, loss of stability with prolonged use often result from the interaction of stabilized magnetic particles with the foreign phase surface. Investigation of the nature and specific features of the interaction of highly dispersed magnetic particles with the surface of different solids permits and prediction of their behavior in each particular case of the use of magnetic fluids.

MATERIALS AND METHODS

The objective of the present study is to establish the main criteria on the change in the stability of magnetic fluids during interaction with foreign surfaces under different conditions.

Magnetic fluids were obtained by peptization of highly dispersed magnetite powder in aqueous and organic media in the presence of surfactants using a conventional method [2]. As model systems, colloid solutions of ferromagnetic metals were used, which have been obtained by electrocondensation [3] in media of different polarity. The foreign solid phase introduced into magnetic fluids was dispersed powders of metal oxides (Al_2O_3, SiO_2, ZnO, PbO), chemical fibers (capron, lavsan, cellulose and flax), as well as metal plates (steel and aluminum). The initial magnetic fluids were stable with respect to spontaneous coagulation particles of disperse particles with time.

Table 1. The main characteristics of dispersion of magnetic particles.

Magnetic particles	Fe_3O_4	Fe	Ni	Co
Average size (diameter), nm	10	12	9	10
Specific surface, m^2 / g	70	57	65	60

The main dispersion characteristics of the magnetic particles under investigation are listed in table 1. The average size (diameter) of metal oxide particles was 1 – 5 μm, the specific surfaces was of the order of 3 m^2 / g. The specific surfaces of chemical fibers and metal plates was of the order of 1 and 0.01 m^2 / g, respectively.

RESULTS AND DISCUSSION

Experimental studies showed that usually the loss of the stability of magnetic fluids during their interaction with foreign solid phase results from the deposition (heterocoagulation) of the magnetic particles on the foreign surface. A characteristic feature of this interaction is the presence of two processes, occurring simultaneously, but at different rates: deposition of magnetic particles (1) on the free surface of material and (2) on the surface aggregates of particles already formed. The contribution of each of these processes to the combined deposition effect depends on the initial ratio of the surfaces of interacting phases (figure 1): $q = S_1 / S_2$, where S_1 and S_2 are the total surfaces of magnetic particles and foreign solid phase. Analysis of the obtained data led to formulation and solution of the kinetic equations of the deposition process [4]. In all cases the proposed theory agrees quite satisfactory with experiment.

To obtain a more detailed concept of the deposition mechanism of magnetic particles, we estimated the influence of the value of S_1 / S_2 on the change of the surface electrical properties of the systems studied. It should be noted that in the majority of cases the surfaces of magnetic particles and foreign solid phase were oppositely charged. It may be expected, therefore, that the interaction of contacting solid phases will follow the neutralization mechanism. To confirm this assumption, we studied experimentally the change of the electrokinetic potential of foreign solid phase with the degree of its surface coverage with magnetic particles. Deposition of magnetic particles leads to recharging of the foreign solid phase surface (figure 2). In most systems the isoelectric state of the surfaces covered by magnetic particles is established in the region of the ratio $S_1 / S_2 = 0.1 - 0.3$.

In the proposed model of the deposition kinetics of magnetic particles the surfaces of contacting solid phases are assumed to be homogeneous. In fact, the surface of actual disperse systems consist of lyophilic and lyophobic patches [5]. This is responsible for the different level of the interaction of the areas of the solid phase surface with the surrounding liquid medium molecules. Therefore, the extent of the loss of stability of magnetic fluids during interaction with foreign solid phase depends, in many respects, on the on the state of their surface boundary layers (solid-liquid). The surface liophility of dispersed phase and their boundary layers is determined by a number of factors: the initial conditions of particle surface, their electrosurface properties, the magnitude of adsorption of surfactants in a solution. Formation of developed boundary layers, including molecules of the medium and surfactants on the surface of magnetic particles and foreign solid surface decre3ases the degree of their interaction [6].

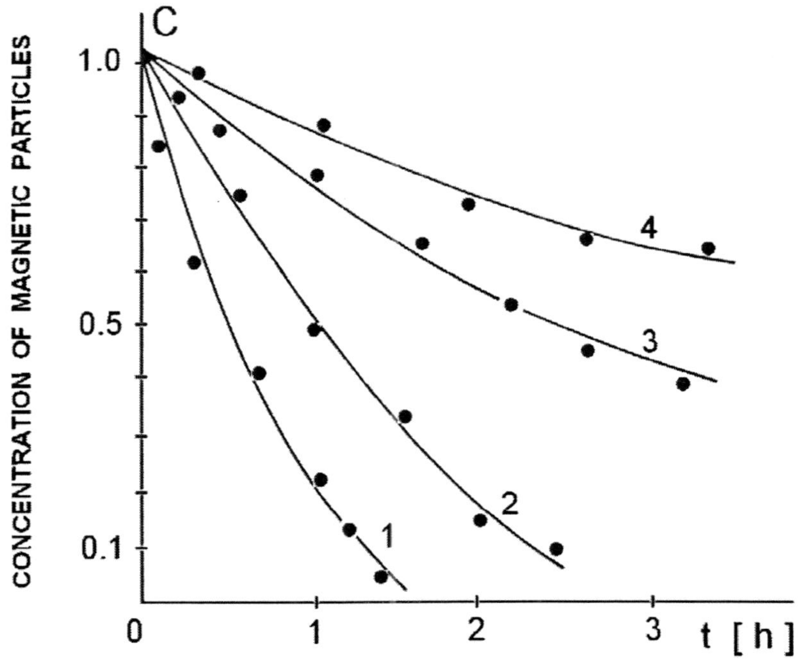

Figure 1. The kinetic of magnetic particles (Fe_3O_4) deposition on the surface of Al_2O_3 in alcohol solution to depends on the parameter $q = S_1 / S_2$: 1 – 0.06; 2 – 0.20; 3 – 1.40; 4 – 4.70.

It was interesting to determine the velocity constant dependence of deposition from the magnitude of liophility of dispersed phase surface. The velocity constant of deposition was calculated according with the process kinetics equation [4] for a case of small magnitude q < 1:

$C = exp\ (- t / qT_1)$ where C is the concentration of magnetic particles, $T_1 \sim 1 / a_1$ is characteristic time of deposition (the magnitude which is in reverse proportion to the velocity constant of the process, a_1). The liophility of solid phase surface was estimated by a comparison of the magnitudes of the electrokinetic potentials of both solid phases to the complex constant of a molecular interaction A_{132} [7]. The values of these parameters characterize the properties of the boundary layers of disperse particles and determine the level of the electrostatic dispersion interaction forces in the systems studied taking into account the contribution of surface-chemical reactions. In the general case, the velocity of deposition decreases with the increase of liophility of the interacting solid phases.

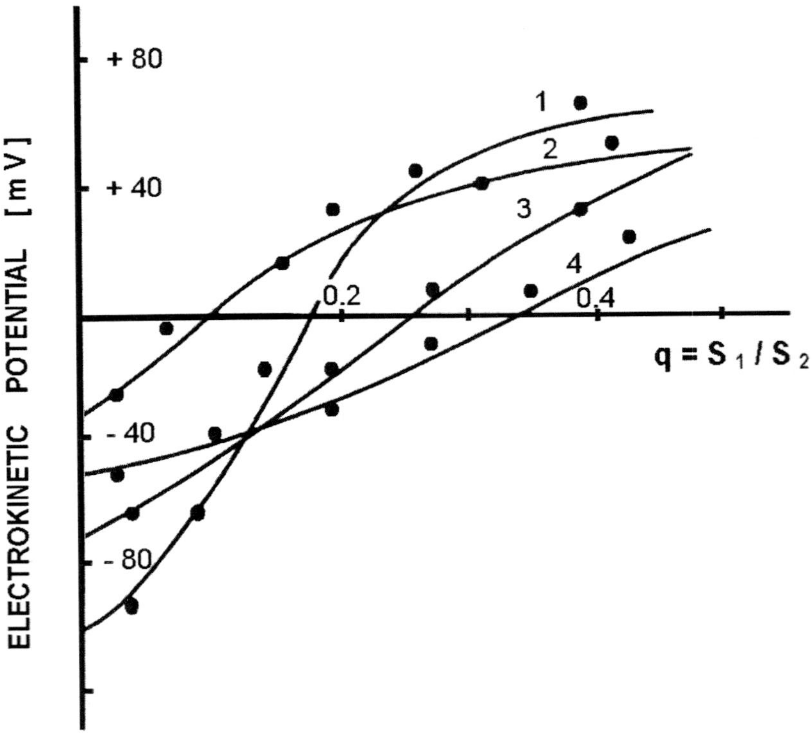

Figure 2. The change of the electrokinetic potential of foreign solid phase with the degree of its surface coverage with magnetic particles in aqueous (1, 3) and alcohol (2, 4) solution: 1 – Fe_3O_4 – SiO_2; 2 – Fe_3O_4 – aluminum plate; 3 – Ni – Al_2O_3; 4 – Co – cellulose.

Conclusion

Analysis of the obtained data reveled the important role of direct surface-chemical interactions during deposition of magnetic particles on different surfaces. Determination of the formation conditions, investigation of the structure and properties of the boundary layers with respect to the nature and composition of dispersion media suggests the ways of controlling the deposition process of magnetic particles.

REFERENCES

[1] S. W. Charles, J. Popplewell. // *Ferromag. Mater.* 2 (1980) 511.

[2] E. E. Bibik. // *Kolloid. Zh.* 35 (1973) 1141.

[3] M. A. Lunina, Yu. A. Novozhilov. // *Kolloid. Zh.* 31 (1969) 120.

[4] M. A. Lunina, V. M. Muller, A. A. Khachaturyan. // *Kolloid. Zh.* 48 (1986) 1190.

[5] E. D. Yakhnin, A. B. Taubman. // *Dokl. Akad. Nauk. USSSR. 152* (1963) 155.

[6] M. A. Lunina, A. A. Khachaturyan, G. F. Paramonova. // *Kolloid. Zh.* 50 (1988) 1197.

[7] J. Visser. // *Adv. Colloid Interf. Sci.* 3 (1972) 335.

In: New Aspects of Biochemical Physics
Editors: Sergei D. Varfolomeev et al., pp. 69-76
ISBN 1-60021-419-3

Chapter 10

CELL MEMBRANE STRUCTURE AND THYROLIBERIN PHYSIOLOGICAL ACTIVITY

V.E. Zhernovkov and N.P. Pal'mina

N.M. Emanual Institute of Biochemical Physics, Russian Academy of Sciences 4 Kosygin str., Moscow 119991, Russia. E-mail: npalm@sky.chph.ras.ru

ABSTRACT

In this study we used an electron spin resonance (EPR) method to explore the effect of thyrotropin-releasing hormone (TRH) in a wide range of concentrations (10^{-4} to 10^{-18} mol/l) on the structure of lipid bilayer of the plasmatic (PM) and endoplasmic reticulum (ER) membranes isolated from mice liver and brain cells. Using 3 stable free radicals localized in different regions of lipids as spin probes we have observed a great changes in lipid microviscosity and order parameter under the effect of different TRH concentrations; the TRH concentration – effect dependence is nonlinear and polymodal. Comparing these changes with the results obtained in in vivo research carried out by I.P.Ashmarin and colleagues, we can conclude that: 1) the joint decreasing of order parameter in ER isolated from liver and brain cells is responsible for the TRH antiepileptic effect; 2) the sharp increasing (decreasing) of lipid microviscosity in PM isolated from liver (brain) under the effect of (10^{-9}-10^{-10}M) TRH is probably connected with interaction with TRH and its receptor; 3) modification of the of biological membrane structure is one of the mechanisms that TRH employs to achieve its anticonvulsant effect. In addition to that our results also support I.P. Ashmarin and colleagues' recommendation that clinically administered doses of TRH should be reduced.

Key words: Thyrotropin-releasing hormone, thyroliberin, endoplasmic reticulum, plasmatic membranes, lipid microviscosity, spin probes, electronic paramagnetic resonance.

Introduction

Thyrotropin-releasing hormone, thyroliberin (TRH), a small neuropeptide (p-Glu-His-Pro-NH_2) initially identified in the hypothalamus, which stimulates the synthesis and secretion of thyroid-stimulating hormone (TSH) and prolactin [1], was shown to be amply distributed in the central nervous system [2]. The presence of TRH was reported not only in the hypothalamus, but also in peripheral tissues including the prostate gland, retina, placenta, gastrointestinal tract, neuronal tissues, male reproductive system and certain endocrine tissues [3].

Experimental research has established that TRH participates in the systemic regulating of a number of vegetative functions such as breathing, the cardiovascular activity, the blood circulation in the brain, maintaining the muscle tone. Yet the bulk of TRH's pharmacological activity range manifests itself in the effect it has on the central nervous system.

TRH has a central stimulating effect that combines with its nootropic activity. In addition to that it acts as an antidepressant. Depending on the concentration, TRH either reduces or cancels out completely the sedative effects that barbiturates, ethanol, benzodiazepines and neuroleptics have on the central nervous system. TRH participates in regulating a number of neural and psychic functions such as the wakefulness level, sleep, emotions, learning abilities, memory.

The most of characteristic feature of TRH is its ability to stimulate respiratory depression as well as to reduce the severity of neurological symptoms in craniocerebral traumas and concussions [5, 6].

The researches carried out by the professor I.P. Ashmarin and colleagues have demonstrated that a broad range of concentrations of TRH (10^{-18} – 10^{-4} M), including ultra low concentrations, could influence on the spontaneous contractive activity of lymphatic vessels and the anti convulsant protection of the brain during epileptic seizures in animals and humans[4,6,7].

One of the key biochemical processes of epilepsy development is the oxidative stress that leads to increased free radical oxidation in the lipids of biological membranes, a decreasing the level of natural antioxidants and to NO generation [8,9]. We have established previously that the vessel contractive activity is directly connected with changes in the key element in the system of lipid peroxide oxidation (LPO), that is the structural property (microviscosity) of the surface areas in the lipid bilayer of the endoplasmatic reticulum (ER) of liver cells [10,11]. However it is a well known that an epileptic seizure is accompanied by an instant emission of large quantities of TRH in the brain [4,5]. Plasmatic membranes (PM) and ER are very important for the development of the process [12,13]. In this context it was of interest to do a comparative in vitro research to study the structure of the different lipid areas of PM and ER isolated from brain and liver cells under the effect of TRH in wide range of concentrations (10^{-18} – 10^{-4} M). It was the aim of our work.

Experimental

We carried out experiments with EPR technique and spin probes employing three stable iminoxyl radicals localizing in different regions of the lipid component of membranes:

2,2,6,6-tetrahydro-4-capryloyl-oxypiperidine-1-oxyl (C7) synthesized at the RAS Institute of Chemical Physics, with NO fragment sinking in the surface regions (2-4 A°) of membrane lipids; 5- and 16-doxylstearic acid produced by the SIGMA company (C5 и C16), localizing at depths of 8 and 20 A° of the membrane, respectively. The final concentration of the spin probes in the membrane suspension was $5x10^{-5}$ M, with a membrane protein concentration of 3,4 mg/ml. Spectrum registration was carried out with Bruker 200D and Bruker EMX EPR spectrometers. The structural changes was estimated as microviscosity or the of the rotary correlation time (τ_C) for C7 and C16 or as the order parameter S – for C5 calculated from the EPR spectrums' readings using the formulas presented in [14]. TRH was stored as 10^{-2}M water solution at 4° C during 10 days. Solutions of TRH at lower concentrations were prepared of an experiment by successive 10 – fold dilution; the volume of the solvent added to a sample was 2 μl.

Endoplasmic reticulum membranes (microsomes) were isolated according to the method [15]; plasmatic membranes – to the method [16]. Protein was determined by the Lowry technique [17].

Results and Discussion

It is known that when TRH is injected into an animal or human organism, it first of all affects the brain, and then the signal spreads down the vessels to other organs and tissues [1-7]. Because of this, our examinations were focused on the PM and ER of the brain as the principal targets of TRH while the similar membranes obtained from the liver were studied as examples of the TRH effect on the contractive activity of blood vessels. We used three probes for mapping the ER membranes of liver cells on which a broad range of TRH concentrations (10^{-10} -10^{-4} M) (figure 1) had been applied. We have found very similar considerable statistically significant structural changes in the lipids. These changes touched upon all the regions of the ER lipid bilayer that were studied. It has to be emphasized that in the concentration range of 10^{-9}-10^{-10}M, TRH caused a statistically significant decrease in microviscosity in both surface regions of the lipids and in those located at deeper levels. It was the dosage range that TRH acted as an anticonvulsant agent in in vivo experiments [7].

Similar experiments on ER membranes from brain cells showed that TRH did not change the microviscosity of the lipid areas at deeper levels, but induced a significant modifications in their order parameter of lipids at the place of probe C5 localization (figure 2 a). High concentrations of TRH (10^{-4}-10^{-5}M) elevated the order parameter S, while concentrations of 10^{-9}-10^{-10} M and 10^{-15}M had the opposite effect. It should be underlined that the concentration – effect dependences of S in the ER membranes isolated from brain and liver were very similar (figure 2). When we compared these results with the data obtained during in vivo experiments in which TRH was injected to mice intranasally [7], we have found that TRH being used in doses decreasing S (10^{-9}-10^{-10} M) delayed the convulsive seizures caused by penthylentetrazol; injected in a concentration 10^{-6}M (S is very close to normal level) had no effect, while concentrations 10^{-4}-10^{-3}M (elevated S) intensified the seizures. As it was mentioned above, ER is regarded as the primary target when degenerative processes develop in the brain [12]. The uniformity of S changes in brain and liver microsomes and the fact that there is a correlation of these changes with in vivo experiments allow us to conclude that it is the structural change in this particular section of ER lipids that is one of the mechanism whereby TRH achieves its anticonvulsant and antiepileptic effect.

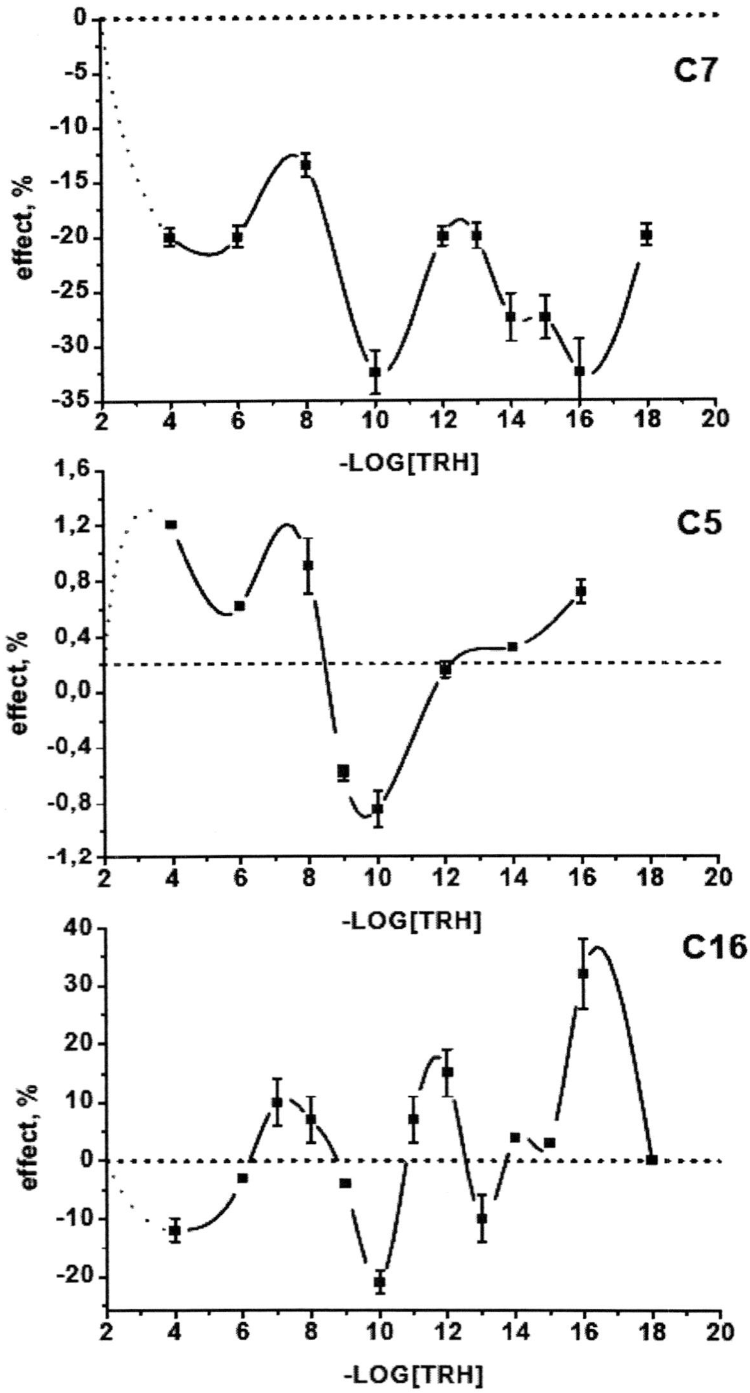

Figure 1. The changes of rotary correlation time τ_C (in relative units) spin probes a) C7 and c) 16DSA; and order parameter S of 5DSA (b) in microsomal membranesfrom liver in dependence on TRH concentration. Spin probe concentration is $5x10^{-5}$ M/L.

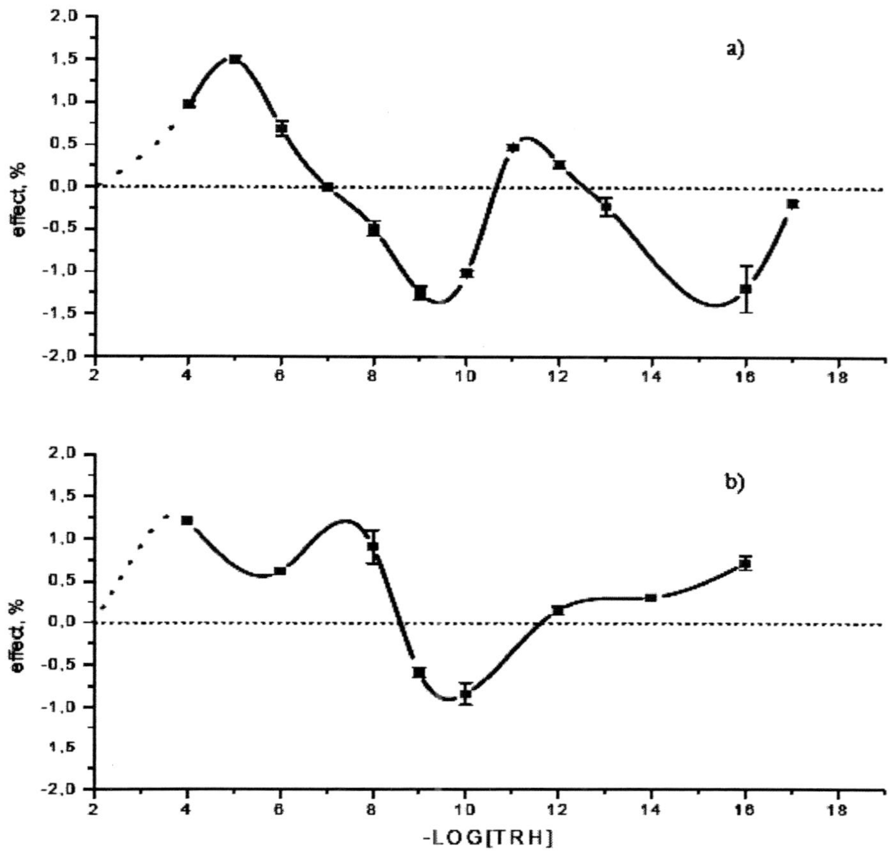

Figure 2. The changes of order parameter S (in relative units) of 5-DSA spin probe in a) endoplasmic reticulum from brain cells and b) endoplasmic reticulum membranes from liver cells in dependence on TRH concentrations. Spin probe concentration is $5x10^{-5}$ M/L.

The specific effect that TRH has on the nervous system is due to its interaction with receptors located on the PM of the brain, with the signal being then passed on by secondary intermediate systems [13]. TRH receptors were found not only in the brain but in a number of other tissues as well: the thymus, the spleen, the pancreas etc. No such receptors were isolated from the liver [13]. We can only assume that the liver contains TRH based on the presence of m-RNA there, a substance necessary for TRH synthesis. The structure of TRH receptors is very well studied. There is a computer model of a TRH receptor that demonstrated that the final stage of the interaction between TRH and the receptor took place in the membrane at a depth of 20A [13]. Since it's in this area that the nitroxyl element of probe C16 is localized, we believed that a comparative study of dependences TRH concentrations - τ_C of this probe in the PM of the brain and liver might be useful for understanding the mechanism of TRH effects. The results presented in figure 3 indicate that, firstly, all the TRH concentrations that were studied increased the microviscosity of the lipids in liver PM and decreased it in brain PM. Secondly, a non linear polymodal association was discovered between the τ_C and the TRH dosage, with 3 concentration intervals that can be identified: high concentration interval (10^{-4} – 10^{-8} M); the interval of receptor concentrations (10^{-9} – 10^{-10} M); ultra low dosages

(10^{-13} M and below). The obvious similarity in the form of the graphs obtained from brain and liver PM indirectly supports the assumption that there have to be TRH receptors in liver PM.

Thus in our experiments TRH caused changes in the structural parameters of all the hydrophobic layers of the lipids of plasmatic membranes and endoplasmatic reticulum isolated from the liver and the brain.

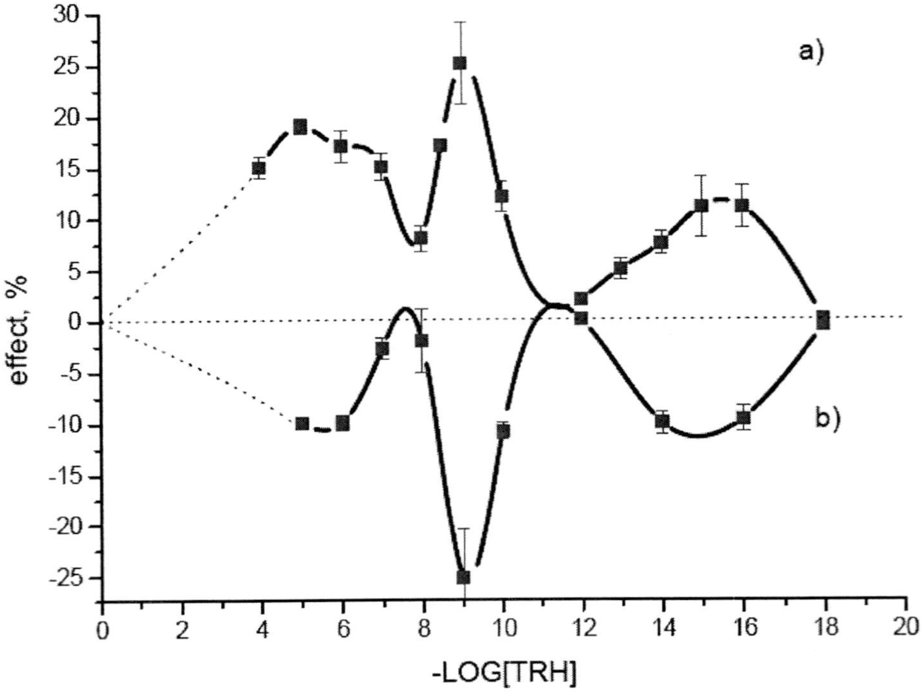

Figure 3. Change in the of rotary correlation time τ_C (in relative units) of spin probe C16 in plasmatic membranes of liver (a) and brain (b) cells in dependence on TRH concentration. Probe concentration of $5x10^{-5}$ M, t=20C.

Conclusions

- TRH in the 10^{-4}-10^{-18}M dosage interval causes non specific total change in the structural properties of the lipids in the endoplasmatic reticulum membranes of liver cells. The TRH concentration – effect dependence is nonlinear and polymodal.
- TRH induces changes in the lipid area of probe C16 (20-22A) location in the plasmatic membranes obtained from the liver and brain, which probably reflects the interaction between TRH and its receptor.
- Comparing the biochemical changes induced by different TRH concentrations observed in our experiments with the results obtained in the research concerning the antiepileptic properties of TRH carried out by I.P.Ashmarin and colleagues using the same concentrations in viro, we can conclude that the modification of the structure of biological membranes is one of the mechanisms that TRH employs to achieve its anticonvulsant effect. In addition to that our results also support I.P. Ashmarin and

colleagues' recommendation that doses of TRH for patient treatment should be reduced in 100 times.

REFERENCES

[1] R.Burgus, T.E. Dunn, D. Desidero, W. Vale, R. Guillemin, *Nature,* Vol.226, N5243 pp.321-325, 1970.

[2] J.E. Morley, *Life Sci.,* Vol.25, N18, pp.1539-1550, 1979.

[3] R. Bilek, *Physiol. Res,* Vol. 49, *Suppl.*1, pp. 19-26, 2000.

[4] I.P Ashmarin, A.P. Kulashev, S.A. Chepurnov , *The USSR Physiological Journal,* Vol. 75, N 5, pp. 627-632, 1989.

[5] A. Horita , *Life Sciences,* Vol. 62, N 17-18, pp. 1443-1448, 1998.

[6] A.N. Inyushkin, N.A., Merkulova, S.A.Chepurnov , *The Russian Physiological Journal.* Vol. 84, N 3, pp.285-292, 1998.

[7] I.P. Ashmarin, L.M. Asanova, K.R. Abbasova, N.E. Chepurnova, G.V. Kosova, S.A. Chepurnov, A.N. Inushkin, O.B. Goncharov, Radiation biology. *Radioecology,* Vol.43, N3, pp.324-327, 2003.

[8] K. Sudha., A.V. Rao, A. Rao, *Clin.Chim.Acta,* ,Vol. 303, N1-2, pp.19-24, 2001.

[9] M.V Frantseva., J.L.P Velazquez., P.A Hwang., P.L Carlen., *Eur. J. Neurosci.,* Vol.12, N4, pp. 1431-1439, 2000.

[10] V.E. Zhernovkov, T.V. Lelekova, N.P. Palmina , Radiation biology. *Radioecology,* Vol.43, N3, pp. 331-333, 2003.

[11] V.E. Zhernovkov., N.P. Palmina, *Biological membranes,* Vol. 22, N5, pp.388-395, 2005.

[12] W. Pashen., *Cell Calcium,* Vol. 34, N4-5, pp. 365-383, 2003.

[13] Y.Sun, X. Lu , M.C. Gershengorn , *J. Molecular Endocrinology,* Vol.30, N2, pp. 87-97, 2003.

[14] A.N. Kuznetsov. Spin probe method. M.: *Science,* 1976. 210 p.

[15] K.Y.Hostetler, D.B.Zenner, H.P. Morris, *Biochim. Biophys. Acta, Vol.*441, № 44. pp. 231 – 238, 1976.

[16] E.G Loten., J.C. Redshaw-Loten, *Analitical Biochem.,. Vol.*154, N 1, pp.183-185, 1986.

[17] O. Lowry, N.Rosenbrouch, A. Barr, R. Randall, *J. Biol. Chem., Vol.* 193, pp. 265-275, 951.

In: New Aspects of Biochemical Physics
Editors: Sergei D. Varfolomeev et al., pp. 77-79
ISBN 1-60021-419-3

Chapter 11

QUANTUM CHEMICAL CALCULATION OF MOLECULE 6-METILPERGIDROTETRALIN

V.A. Babkin, R.G. Fedunov, A.A. Ostrouhov, A.V. Kudryashov, G.E. Zaikov and T.V. Peresypkina

119991, Moscow, Russia, N.M. Emanuel Institute of Biochemical Physics,
4 Kosygin Street. E-mail: chembio@sky.chph.ras.ru;
403343, SF VolgGASU, c. Mikhailovka, region Volgograd s. Michurina 21.
E-mails: sfi@reg.avtlg.ru

ABSTRACT

For the first time it is executed quantum chemical calculation of a molecule 6-metilpergidrotetralin by method AB INITIO in basis 6-311 G ** with optimization of geometry on all parameters. The optimized geometrical and electronic structure of this connection is received. Acid force 6-metilpergidrotetralin is theoretically appreciated. It is established, that it concerns to a class of weak H-acids (pKa = + 35.85, pKa - a universal index of acidity), as pKa> 14.

AIMS AND BACKGROUND

The purpose of the present work is quantum chemical calculation of a molecule 6-metilpergidrotetralin method AB INITIO in basis 6-311 G ** [1] with optimization of geometry on all parameters in approximation of an isolated molecule a gas phase and an estimation of his acid force.

RESULTS OF CALCULATIONS

In connection with that the molecule 6-metilpergidrotetralin exists in two configurations of a ring "bed" and "armchair" calculation was carried out by method AB INITIO in basis 6-

311 G ** both configurations. It is shown, that the configuration "bed" is less energetically favorable (~ on 41.8 kDg/mol'). Therefore the analysis of quantum chemical parameters and an estimation of acid force were carried out for a configuration "armchair".

Geometrical and electronic structure, general energy of 6-metilpergidrotetralin (a configuration "armchair")–E_0,total energy of connections–E_{con} . Are shown on figure 1. Using the known formula [2] pKa = 49.04-134.61 * $q_{H^+}^{max}$ (AB INITIO in basis 6-311 G ** R = 0,97, R- coefficient of correlation) for a researched molecule 6-метилпергидротетралина (q_{H^+} = +0.098, see rice 1) is determined pKa = + 35.85

Conclusions

Thus, the executed quantum chemical calculation of a molecule 6-metilpergidrotetralin by method AB INITIO in basis 6-311 G ** has allowed to receive the optimized electronic and geometrical structure of this connection. The theoretical estimation of acid force 6-metilpergidrotetralin pKa = + 35.85 has shown, that it concerns to a class of very weak H-acids (pKa> +14)

Lengths of Connections and Valent Corners of a Molecule 6-Metilpergidrotetralin

R (C-C) = 0.1532 nm, R (C1-C) = 0.1532 nm, R (C2-C) = 0.1543 nm,

R (C-H) = 0.1088 nm, R (C1-H) = 0.1093 nm, R (C3-H) = 0.1083 nm,

∠(C-C-C) = 111.8^0, ∠ (C-C-C1) = 111.2^0, ∠ (C-C1-C2) = 112.5^0,
∠ (C-C2-C3) = 109.2^0,

∠ (C1-C2-C3) = 113.1^0, ∠ (C-C-C2) = 113.7^0,

∠ (C-C-H) = 109.6^0, ∠ (C2-C1-H) = 105.5^0, ∠ (C2-C3-H) = 109.6^0,
∠ (C-C1-H) = 106.2^0,

∠ (H-C-H) = 106.8^0.

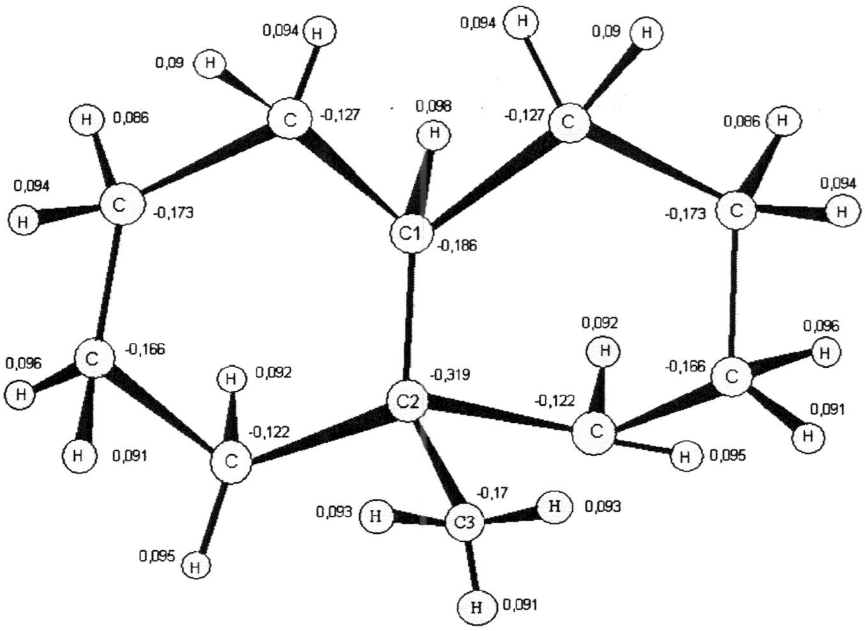

Figure. Rice 1 the Geometrical and electronic structure 6-metilpergidrotetralin, conformation rings - "armchair" (E_0 =-1123445 kDg/mol', D = 0.039 dB.).

References

[1] M.W. Schmidt, K.K. Baldridge, J.A. Elbert, M.S. Gordon, J.H. Ensen, S.Koseki, N. Matsunaga, K.A. Nguyen, S.J. Su, T.L. Windus, Together with M. Dupuis, J.A. Montgomery, *J. Comput. Chem.* 14, 1347-1363(1993).

[2] Babkin V.A., R.G. Fedunov, K.S. Minsker, O.A. Ponomarev, Y. A. Sangalov, A.A. Berlin, G.E. Zaikov, Connection of the universal acidity index of H-acids with the charge on Hydrogen atom (AB INITIO METHOD). *Oxidation Communication,* 2002, 25, №1, pp. 21 - 47.

In: New Aspects of Biochemical Physics
Editors: Sergei D. Varfolomeev et al., pp. 81-83
ISBN 1-60021-419-3

Chapter 12

QUANTUM CHEMICAL CALCULATION OF MOLECULE 7-METILPERGIDROTETRALIN

V.A. Babkin, R.G. Fedunov, A.A. Ostrouhov, R.A. Reshetnikov and G.E. Zaikov

119991, Moscow, Russia, N.M. Emanuel Institute of Biochemical Physics,
4 Kosygin Street. E-mail: chembio@sky.chph.ras.ru
403343, SF VolgGASU, c. Mikhailovka, region Volgograd
s. Michurina 21. E-mails: sfi@reg.avtlg.ru

ABSTRACT

For the first time it is executed quantum chemical calculation of a molecule 7-metilpergidrotetralin by method AB INITIO in basis 6-311 G ** with optimization of geometry on all parameters. The optimized geometrical and electronic structure of this connection is received. Acid force 7-metilpergidrotetralin is theoretically appreciated. It is established, that it concerns to a class of weak H-acids (pKa = + 35.31, pKa - a universal index of acidity), as pKa> 14.

AIMS AND BACKGROUND

The purpose of the present work are quantumchemical calculation of a molecule 7-metilpergidrotetralin method AB INITIO in basis 6-311 G ** [1] with optimization of geometry on all parameters in approximation of an isolated molecule a gas phase and an estimation of his acid force.

RESULTS OF CALCULATIONS

In connection with that the molecule of 7-metilpergidrotetralin exists in two configurations of a ring "bed" and "armchair" calculation it was carried out by method AB

INITIO in basis 6-311 G ** both configurations. It is shown, that the configuration "bed" is less energetically favorable (~ on 41.8 kDg/mol'). Therefore the analysis of quantum chemical parameters and an estimation of acid force were carried out for a configuration "armchair".

Geometrical and electronic structure, general energy 7-metilpergidrotetralin (a configuration "armchair")-E_0, total energy of connections- E_{con}. Are shown on figure 1. Using the known formula [2] pKa = 49.04-134.61 * q_{H+}^{max} (AB INITIO in basis 6-311 G ** R = 0,97, R-coefficient of correlation) for a researched molecule of 7-metilpergidrotetralin (q_{H+} = +0,01, see rice 1) pKa = + 35.31 is determined.

Conclusions

Thus, the executed quantum chemical calculation of a molecule metilpergidrotetralin by method AB INITIO in basis 6-311 G ** has allowed to receive the optimized electronic and geometrical structure of this connection. The theoretical estimation of acid force 7-metilpergidrotetralin pKa = + 35.31 has shown, that it concerns to a class of very weak H-acids (pKa> +14)

Lengths of Connections and Valent Corners of a Molecule 7-Metilpergidrotetralin

R (C - C) =0.1532 nm, R (C_1 - C) =0.1532 nm,

R (C_2 - C) =0.1544 nm, R (C - H) =0.1088 nm,

R (C_1 - H) =0.1093 nm, R (C_3 - H) =0.1087 nm,

$\angle$ (C – C – C)=111.2^0, $\angle$ (C – C – C_1)=112.3^0,

$\angle$ (C_1 – C – C_2)=111.1^0, $\angle$ (C – C_2 – C_3)=113.3^0, $\angle$ (C – C – H)=109.8^0,

$\angle$ (C – C_1 – H)=107.4^0,

$\angle$ (C_2 – C_3 – H)=111.0^0, $\angle$ (C – C_2 – H)=107.5^0, $\angle$ (H – C – H)=106.8^0.

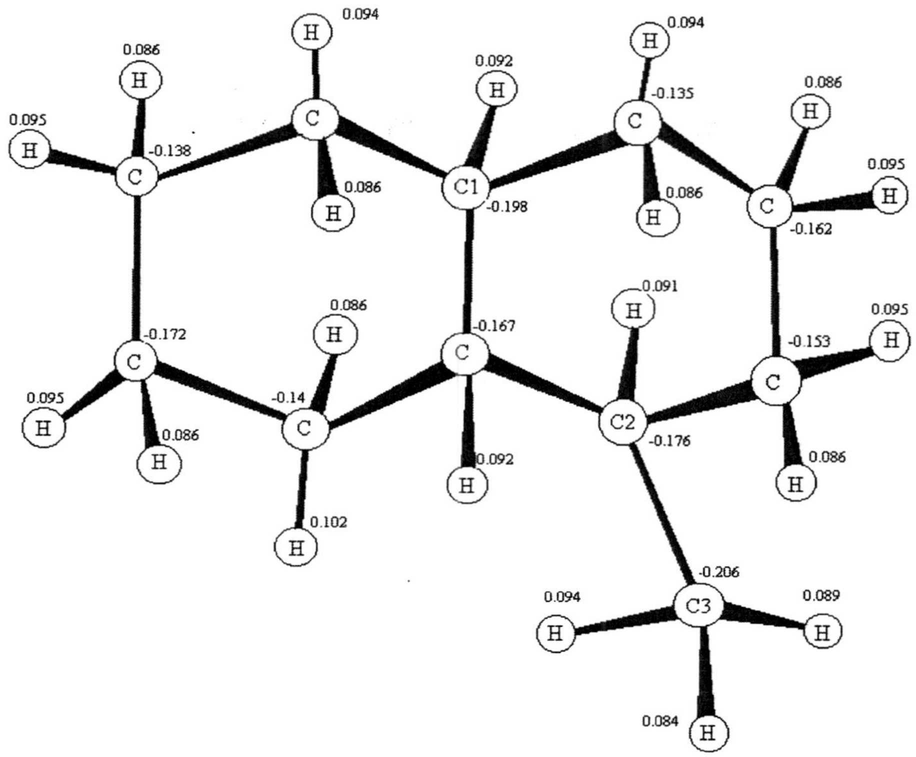

Figure. Rice 1 the Geometrical and electronic structure 7-metilpergidrotetralin, conformation rings - "armchair". (E_0 = - 1123467.977 kDg/mol'; D=0.075 dB.).

REFERENCES

[1] M.W. Schmidt, K.K. Baldridge, J.A. Elbert, M.S. Gordon, J.H. Ensen, S.Koseki, N. Matsunaga, K.A. Nguyen, S.J. Su, T.L. Windus, Together with M. Dupuis, J.A. Montgomery, *J. Comput. Chem.* 14, 1347-1363(1993).

[2] Babkin V.A., R.G. Fedunov, K.S. Minsker, O.A. Ponomarev, Y. A. Sangalov, A.A. Berlin, G.E. Zaikov, Connection of the universal acidity index of H-acids with the charge on Hydrogen atom (AB INITIO METHOD). *Oxidation Communication,* 2002, 25, №1, pp. 21 - 47.

In: New Aspects of Biochemical Physics
Editors: Sergei D. Varfolomeev et al., pp. 85-88
ISBN 1-60021-419-3

Chapter 13

About a Geometrical and Electronic Structure of a Molecule Gopan

V.A. Babkin, R.G. Fedunov, A.A. Ostrouhov and G.E.Zaikov
119991, Moscow, Russia, N.M. Emanuel Institute of Biochemical Physics,
4 Kosygin Street. E-mail: chembio@sky.chph.ras.ru
403343, SF VolgGASU, c. Mikhailovka, region Volgograd s. Michurina 21.
E-mails: sfi@reg.avtlg.ru

Abstract

For the first time it is executed quantum chemical calculation of a molecule of gopan, being one of the major intermediate products in the most complicated evolutionary process of natural synthesis of oil, method AB INITIO in basis 6-311 G ** with optimization of geometry on all parameters. The optimized geometrical and electronic structure of this connection is received. Acid force gopan is theoretically appreciated. It is established, that it concerns to a class of very weak H-acids (pKa = + 33, a pKa-universal index of acidity), since pKa> 14.

Aims and Background

The purpose of the present work is quantum chemical calculation of a molecule of gopan, being one of the major intermediate products in the most complicated evolutionary process of natural synthesis of oil, method AB INITIO in basis 6-311 G ** [1] with optimization of geometry on all parameters by a gradient method, with accuracy of calculation of a gradient up to 0.0001 in approximation of an isolated molecule in a gas phase and a theoretical estimation of its acid force.

RESULTS OF CALCULATIONS

The schematic, geometrical and electronic structure, full energy of system E_0 =-3059847.437 kDg/mol', dipol moment D=0.0922 dB, lengths of bonds - R, valent θ and torses corners φ, atoms are adhered A1, A2, A3 and charges on atoms - q of a molecule gopan are shown on figure 1 and in table 1. Using the known formula pKa=49.04-134.61 · q^{max}_{H+} [2] (for method AB INITIO in basis 6-311 G **, R=0.97, the R- coefficient of correlation) for a researched molecule of gopan (q^{max}_{H+} = + 0.119, see table 1 q_{H51}) pKa = + 33 is determined.

CONCLUSIONS

Thus, for the first time executed quantum chemical calculation of a molecule of gopan method AB INITIO in basis 6-311 G ** has allowed to receive optimized on all parameters by a gradient method the Geometrical and electronic structure of this connection. The theoretical estimation of acid force of a molecule of gopan pKa =+ 33 has shown, that it concerns to a class of very weak H-acids (pKa> +14).

Full energy of system E_0 =-3059847.437 kDg/mol'; dipol the moment-D=0.0922 dB; length of bonds - R; a valent corner θ; torses corners φ; atoms of a binding - A1, A2, A3; a q-charge on atom of a molecule gopan.

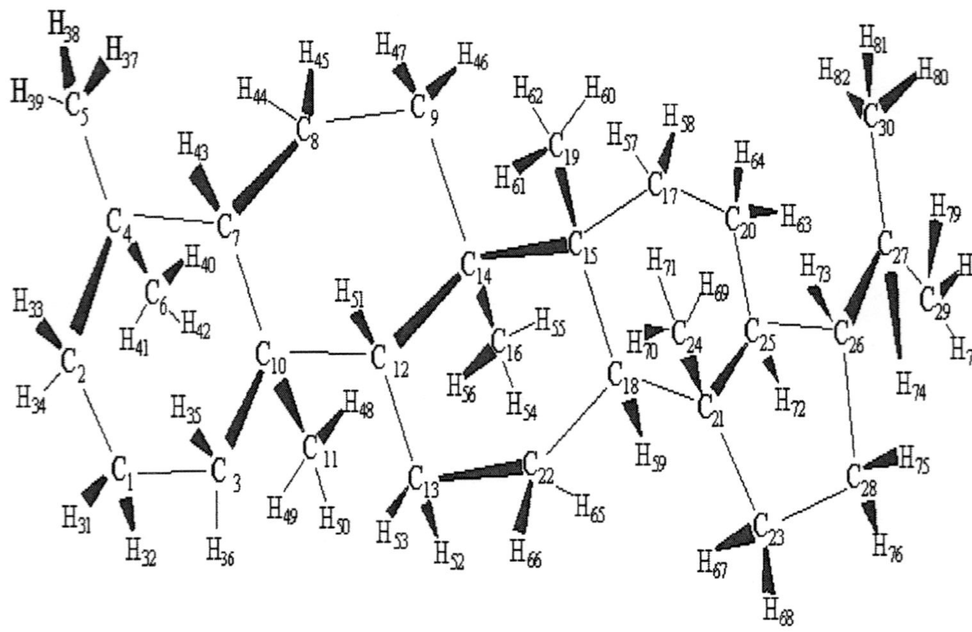

Figure 1. Schematic structure of the molecule gopan.

Table 1.

№	Atom	R,A	θ,^{0}	φ,^{0}	A1	A2	A3	q
1	C	-	-	-	-	-	-	-0.14131
2	C	1.5244	-	-	1	-	-	-0.13954
3	C	1.5296	111.782	-	1	2	-	-0.13761
4	C	1.5425	114.259	55.131	2	1	3	-0.25781
5	C	1.5428	106.641	191.204	4	2	1	-0.18497
6	C	1.5402	110.561	75.377	4	2	1	-0.17885
7	C	1.5636	108.304	308.873	4	2	1	-0.15617
8	C	1.5332	114.543	183.003	7	4	2	-0.09636
9	C	1.5309	111.266	161.936	8	7	4	-0.10382
10	C	1.5502	114.051	304.747	3	1	2	-0.24007
11	C	1.5476	107.424	287.597	10	3	1	-0.15893
12	C	1.5779	107.938	165.706	10	3	1	-0.12848
13	C	1.5427	113.952	59.931	12	10	3	-0.11071
14	C	1.5485	114.599	54.556	9	8	7	-0.25766
15	C	1.6056	110.790	196.791	14	9	8	-0.26434
16	C	1.5534	105.744	76.131	14	9	8	-0.16394
17	C	1.5563	111.241	305.630	15	14	9	-0.11717
18	C	1.5695	106.776	184.790	15	14	9	-0.10556
19	C	15568	110.818	62.784	15	14	9	-0.16628
20	C	1.5402	115.329	197.408	17	15	14	-0.08684
21	C	1.5507	115656	165.523	18	15	14	-0.29667
22	C	1.5339	113.562	187.529	13	12	10	-0.10391
23	C	1.5393	113.834	196.637	21	18	15	-0.10155
24	C	1.5446	114.404	73.550	21	18	15	-0.15731
25	C	1.5264	109.024	54.148	20	17	15	-0.16041
26	C	1.5449	121.930	171.936	25	20	17	-0.15850
27	C	1.5452	116.863	279.115	26	25	20	-0.16653
28	C	1.5438	104.328	155.570	23	21	18	-0.14543
29	C	1.5350	110.425	176.461	27	26	25	-0.21206
30	C	1.5321	114.980	52.557	27	26	25	-0.22107
31	H	1.0877	109.304	175.704	1	2	4	0.09646
32	H	1.0858	110.311	291.522	1	2	4	0.08901
33	H	1.0897	108.307	294.427	2	1	3	0.09069
34	H	1.0882	110.093	178.621	2	1	3	0.09551
35	H	1.0894	108.199	66.243	3	1	2	0.08962
36	H	1.0850	108.648	181.550	3	1	2	0.10208
37	H	1.0843	112.544	189.907	5	4	2	0.09116
38	H	1.0868	111.502	68.273	5	4	2	0.08972
39	H	1.0870	109.836	309.109	5	4	2	0.09074
40	H	1.0856	111.395	182.283	6	4	2	0.09308
41	H	1.0872	109.505	63.860	6	4	2	0.08888
42	H	1.0806	113.472	304.042	6	4	2	0.09032
43	H	1.0914	103.202	297.258	7	4	2	0.10260
44	H	1.0840	110.337	42.837	8	7	4	0.10035
45	H	1.0840	110.934	286.058	8	7	4	0.09127
46	H	1.0867	109.003	177.047	9	8	7	0.09879
47	H	1.0824	106.809	291.124	9	8	7	0.09602
48	H	1.0800	113.759	158.980	11	10	3	0.09066
49	H	1.0807	111.359	38.097	11	10	3	0.08996
50	H	1.0828	110.754	280.099	11	10	3	0.09392
51	H	1.0845	102.716	306.002	12	10	3	0.11900
52	H	1.0836	110.787	65.129	13	12	10	0.09208
53	H	1.0836	109.886	308.464	13	12	10	0.09948

Table 1. (Continued).

54	H	1.0798	114.135	198.847	16	14	9	0.09179
55	H	1.0817	111.567	77.210	16	14	9	0.09307
56	H	1.0800	110.626	319.259	16	14	9	0.09279
57	H	1.0864	108.800	74.529	17	15	14	0.10117
58	H	1.0830	110.808	318.390	17	15	14	0.09593
59	H	1.0843	105.455	52.650	18	15	14	0.11509
60	H	1.0807	110.703	205.803	19	15	14	0.09257
61	H	1.0800	114.044	85.111	19	15	14	0.09421
62	H	1.0808	111.811	323.657	19	15	14	0.09354
63	H	1.0825	107.874	176.009	20	17	15	0.10295
64	H	1.0857	111.624	291.818	20	17	15	0.09527
65	H	1.0870	108.386	189.454	22	13	12	0.09303
66	H	1.0841	109.215	74.624	22	13	12	0.08837
67	H	1.0876	109.858	37.725	23	21	18	0.09315
68	H	1.0853	113.490	277.812	23	21	18	0.09845
69	H	1.0838	111.046	176.227	24	21	18	0.09031
70	H	1.0851	110.019	58.305	24	21	18	0.09433
71	H	1.0782	114.177	297.350	24	21	18	0.09240
72	H	1.0926	106.450	50.700	25	20	17	0.09994
73	H	1.0891	109.120	41.152	26	25	20	0.09871
74	H	1.0907	107.132	293.575	27	26	25	0.08789
75	H	1.0847	110.984	97.830	28	23	21	0.10311
76	H	1.0845	110.404	215.363	28	23	21	0.09951
77	H	1.0865	110.799	174.914	29	27	26	0.08929
78	H	1.0844	112.366	54.495	29	27	26	0.09091
79	H	1.0882	110.570	294.131	29	27	26	0.08314
80	H	1.0867	110.005	174.524	30	27	26	0.08970
81	H	1.0877	111.488	55.311	30	27	26	0.08636
82	H	1.0834	112.159	293.808	30	27	26	0.09450

References

[1] M.W. Schmidt, K.K. Baldridge, J.A. Elbert, M.S. Gordon J.H. Ensen, S. Koseki, N. Matsunaga, K.A. Nguyen, S.J. Su, T.L. Winds, together with M. Dupuis, J.A. Montgomery. *J. Comput. Chem.,* 1993 14, pp.1347-1363.

[2] V.A. Babkin, R.G. Fedunov, K.S. Minsker, O.A. Ponomarev, Y.A. Sangalov, A.A. Berlin, G.E. Zaikov. Connection of the universal acidity index of H-acids with the charge on hydrogen atom (AB INITIO METHOD). *Oxidation Communication,* 2002, №1, 25, pp. 21-47.

In: New Aspects of Biochemical Physics
Editors: Sergei D. Varfolomeev et al., pp. 89-93 ISBN 1-60021-419-3

Chapter 14

About a Geometrical and Electronic Structure of a Molecule Diagopan

V.A. Babkin, R.G. Fedunov, A.A. Ostrouhov and G.E. Zaikov
119991, Moscow, Russia, N.M. Emanuel Institute of Biochemical Physics,
4 Kosygin Street. E-mail: chembio@sky.chph.ras.ru
403343, SF VolgGASU, c. Mikhailovka, region Volgograd
s. Michurina 21. E-mails: sfi@reg.avtlg.ru

Abstract

For the first time it is executed quantum chemical calculation of a molecule of diagopan, being one of the major intermediate products in the most complicated evolutionary process of natural synthesis of oil, method AB INITIO in basis 6-311 G ** with optimization of geometry on all parameters. The optimized geometrical and electronic structure of this connection is received. Acid force diagopan is theoretically appreciated. It is established, that it concerns to a class of very weak H-acids (pKa = + 34.7, a pKa-universal index of acidity), since pKa> 14.

Aims and Background

The purpose of the present work is quantum chemical calculation of a molecule diagopan, being one of the major intermediate products in the most complicated evolutionary process of natural synthesis of oil, method AB INITIO in basis 6-311 G ** [1] with optimization of geometry on all parameters by a gradient method, with accuracy of calculation of a gradient up to 0.0001 in approximation of an isolated molecule in a gas phase and a theoretical estimation of its acid force.

RESULTS OF CALCULATIONS

The schematic, geometrical and electronic structure, full energy of system E_0 =-3059885.47 kDg/mol', dipol moment D=0.1968 dB, lengths of bonds - R, valent θ and torses corners φ, atoms are adhered A1, A2, A3 and charges on atoms - q of a molecule diagopan are shown on figure 1 and in table 1. Using the known formula pKa=49.4-134.61 · q^{max}_{H+} [2] (for method AB INITIO in basis 6-311 G **, R=0.97, the R- coefficient of correlation) for a researched molecule of digopan (q^{max}_{H+} = + 0.10655, see table 1 q_{H59}) pKa = + 34,7 is determined.

CONCLUSIONS

Thus, for the first time executed quantum chemical calculation of a molecule of diagopan by method AB INITIO in basis 6-311 G ** has allowed to receive optimized on all parameters by a gradient method a geometrical and electronic structure of this connection. The theoretical estimation of acid force of a molecule of diagopan pKa = + 34.7 has shown, that it concerns to a class of very weak H-acids (pKa> +14).

Full energy of system E_0 =-3059885.47 kDg/mol'; dipole the moment-D=0.1968 dB; length of bonds - R; a valent corner θ; torses corners φ; atoms of a binding - A1, A2, A3; a q-charge on atom of a molecule diagopan.

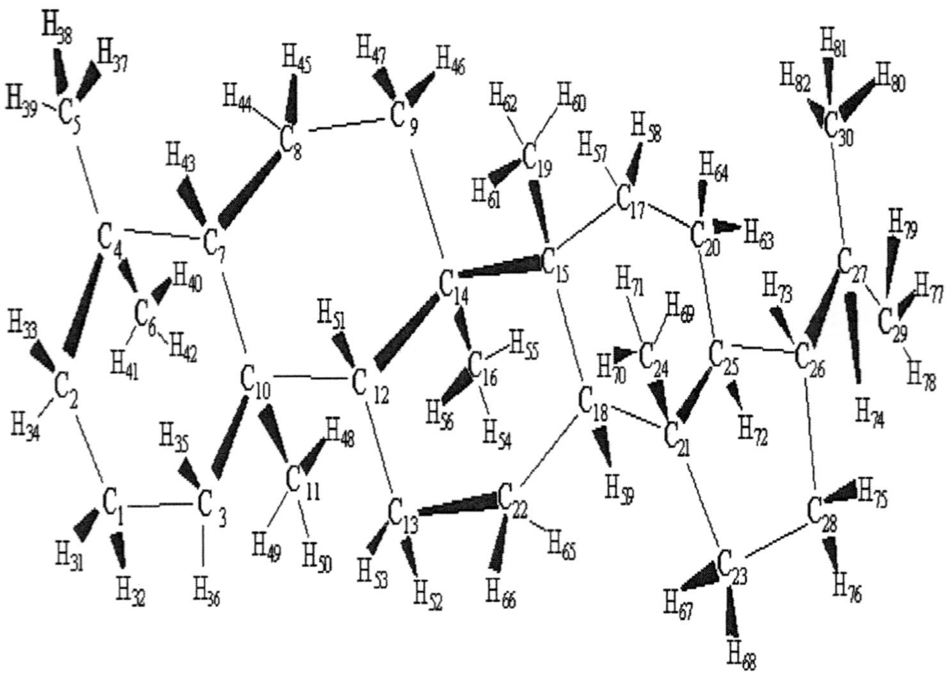

Figure 1. Schematic structure of the molecule diagopan.

Table 1.

№	Атом	R,A	θ	φ	A1	A2	A3	q
1	C	-	-	-	-	-	-	-0.14253
2	C	1,5243	-	-	1	-	-	-0.13808
3	C	1,5302	112.012	-	1	2	-	-0.13823
4	C	1,5421	114.126	55.007	2	1	3	-0.25872
5	C	1,5426	106.742	191.337	4	2	1	-0.18564
6	C	1,5398	110.621	75.376	4	2	1	-0.17833
7	C	1,5623	108.120	308.892	4	2	1	-0.15157
8	C	1,5302	114.829	183.568	7	4	2	-0.09160
9	C	1,5287	111.040	162.655	8	7	4	-0.12442
10	C	1,5503	114.009	304.902	3	1	2	-0.24622
11	C	1,5471	107.555	287.287	10	3	1	-0.15860
12	C	1,5775	108.321	165.869	10	3	1	-0.13918
13	C	1,536	113.719	57.263	12	10	3	-0.09798
14	C	1,5417	115.266	57.011	9	8	7	-0.26018
15	C	1,5905	110.495	195.469	14	9	8	-0.10856
16	C	1,548	107.847	74.804	14	9	8	-0.16224
17	C	1,5748	115.471	308.408	15	14	9	-0.16234
18	C	1,5609	108.571	178.355	15	14	9	-0.13188
19	C	1,552	115.672	203.872	17	15	14	-0.08165
20	C	1,5387	114.072	170.300	18	15	14	-0.32900
21	C	1,5237	111.421	191.222	13	12	10	-0.10659
22	C	1,5367	114.870	190.895	20	18	15	-0.09151
23	C	1,5444	112.042	66.357	20	18	15	-0.16777
24	C	1,5159	113.495	38.920	19	17	15	-0.15101
25	C	1,5469	119.456	179.105	24	19	17	-0.15361
26	C	1,5445	116.288	272.479	25	24	19	-0.16593
27	C	1,5407	104.177	156.666	22	20	18	-0.14498
28	C	1,5346	110.488	174.213	26	25	24	-0.21081
29	C	1,5321	114.562	50.449	26	25	24	-0.22317
30	C	1,542	113.985	82.211	17	15	14	-0.20523
31	H	1,0876	109.325	175.659	1	2	4	0.09665
32	H	1,0859	110.241	291.423	1	2	4	0.08892
33	H	1,0897	108.374	294.311	2	1	3	0.09085
34	H	1,0882	110.075	178.481	2	1	3	0.09532
35	H	1,0895	108.325	66.455	3	1	2	0.09110
36	H	1,0846	108.405	181.734	3	1	2	0.10188
37	H	1,0844	112.500	190.102	5	4	2	0.09128
38	H	1,0868	111.493	68.514	5	4	2	0.08699
39	H	1,0870	109.853	309.326	5	4	2	0.09080
40	H	1,0856	111.381	182.953	6	4	2	0.09291
41	H	1,0872	109.512	64.529	6	4	2	0.08909

Table 1. (Continued).

42	H	1,0807	113.457	304.703	6	4	2	0.08997
43	H	1,0912	103.107	297.742	7	4	2	0.10366
44	H	1,0841	110.374	43.776	8	7	4	0.10083
45	H	1,0843	111.171	286.696	8	7	4	0.09051
46	H	1,0849	108.561	180.316	9	8	7	0.10297
47	H	1,0865	107.225	295.481	9	8	7	0.09863
48	H	1,0801	113.661	159.275	11	10	3	0.09125
49	H	1,0808	111.460	38.310	11	10	3	0.09016
50	H	1,0832	110.614	280.232	11	10	3	0.09405
51	H	1,0900	103.530	303.554	12	10	3	0.10543
52	H	1,0850	111.332	69.243	13	12	10	0.09062
53	H	1,0827	110.295	311.623	13	12	10	0.10073
54	H	1,0885	105.275	64.668	15	14	9	0.10653
55	H	1,0812	113.614	199.492	16	14	9	0.09216
56	H	1,0840	110.680	78.922	16	14	9	0.09208
57	H	1,0808	111.535	320.546	16	14	9	0.09096
58	H	1,0864	108.124	322.440	17	15	14	0.10245
59	H	1,0869	107.778	54.440	18	15	14	0.10655
60	H	1,0868	109.594	276.704	19	17	15	0.09155
61	H	1,0835	107.024	162.694	19	17	15	0.10516
62	H	1,0868	109.232	182.850	21	13	12	0.09321
63	H	1,0877	108.888	66.787	21	13	12	0.08929
64	H	1,0883	109.503	39.179	22	20	18	0.09382
65	H	1,0851	113.845	279.221	22	20	18	0.09798
66	H	1,0845	111.866	191.307	23	20	18	0.09418
67	H	1,0850	110.197	72.314	23	20	18	0.09622
68	H	1,0829	112.857	312.323	23	20	18	0.09360
69	H	1,0920	108.103	56.612	24	19	17	1.10076
70	H	1,0892	108.823	33.654	25	24	19	0.09851
71	H	1,0909	107.222	291.573	26	25	24	0.08833
72	H	1,0844	110.782	211.379	27	22	20	0.09831
73	H	1,0851	110.626	93.711	27	22	20	0.10127
74	H	1,0865	110.920	174.993	28	26	25	0.08941
75	H	1,0846	112.327	54.507	28	26	25	0.09027
76	H	1,0882	110.483	294.268	28	26	25	0.08303
77	H	1,0867	110.094	176.503	29	26	25	0.09033
78	H	1,0876	111.369	57.327	29	26	25	0.08656
79	H	1,0839	112.161	295.986	29	26	25	0.09450
80	H	1,0777	114.592	291.926	30	17	15	0.09650
81	H	1,0870	109.384	171.716	30	17	15	0.08543
82	H	1,0880	110.483	53.995	30	17	15	0.08400

References

[1] M.W. Schmidt, K.K. Baldridge, J.A. Elbert, M.S. Gordon J.H. Ensen, S. Koseki, N. Matsunaga, K.A. Nguyen, S.J. Su, T.L. Winds, together with M. Dupuis, J.A. Montgomery. *J. Comput. Chem.,* 14,1347-1363(1993).

[2] Babkin V.A.., Fedunov R.G., Minsker K.S., Ponomarev O.A.,Sangalov Y.A., Berlin A.A., Zaikov G.E. Connection of the universal acidity index of H-acids with the charge on hydrogen atom (AB INITIO METHOD). *Oxidation Communication,* 2002, №1, 25, p.p. 21-47.

In: New Aspects of Biochemical Physics
Editors: Sergei D. Varfolomeev et al., pp. 95-112 ISBN 1-60021-419-3

Chapter 15

SELECTIVE OXIDATION OF ETHYLBENZENE WITH DIOXYGEN INTO α-PHENYLETHYLHYDROPEROXIDE. MODIFICATION OF CATALYST ACTIVITY OF NI(II) AND FE(III) COMPLEXES UPON ADDITION OF QUATERNARY AMMONIUM SALTS AS EXOLIGANDS

L.I. Matienko and L.A. Mosolova
N.M. Emanuel's Institute of Biochemical Physics of Russian Academy of Sciences, 4 Kosygin str., Moscow 119991, Russia; E-mail:matienko@sky.shph.ras.ru

ABSTRACT

Based on the established mechanism of catalysis by nickel complexes $Ni(II)(L^1)_2$ (L^1=acac$^-$), activated by additives of electron-donor monodentate ligands-modifiers L^2 (L^2=HMFA, DMFA, MP, MSt) in the process of selective alkylarens (ethylbenzene and cumene) oxidation with dioxygen into corresponding hydroperoxides (PEH) the effective methods for control of the selective ethylbenzene oxidation into α-phenylethylhydroperoxide (PEH), including the use of quaternary ammonium salts R_4NBr as L^2 exoligands-modifiers are suggested by us. Values of selectivity, S_{PEH}, conversion degree, C, and PEH yield reached at application of L^2=R_4NBr (Me_4NBr) exceed analogous parameters in the presence of the other $\{Ni(II)(L^1)_2+L^2\}$ systems and known catalysts of ethylbenzene oxidation into PEH. The principle possibility of the catalysts construction on the basis of complexes $Fe(III)(acac)_3$ and R_4NBr for the increase in rate and selectivity of the ethylbenzene oxidation by dioxygen into PEH is demonstrated. The proposed "dioxygenase-like" mechanism of the transformation of iron catalyst into more active catalytic particles in the course of oxidation process is discussed. The value method of activity of nickel and iron catalysts in the micro steps of the chain-radical ethylbenzene oxidation is offered.

Key words: oxidation, ethylbenzene, α-phenylethylhydroperoxide, homogeneous catalysis, dioxygen, nickel (II) bis(acetylacetonate), iron (III) tris(acetylacetonate), quaternary ammonium salts.

INTRODUCTION

In spite of theoretical interest the problem of selective oxidation of alkylarens (ethylbenzene and cumene) in ROOH, primary products of RH oxidation, is of current importance from practical point of view in connection with ROOH use in large-tonnage productions such as production of propylene oxide and styrene (α-phenylethylhydroperoxide), or phenol and acetone (cumyl hydroperoxide) [1].

Except catalytic systems developed by us nobody had proposed effective catalysts for selective oxidation of ethylbenzene into α-phenylethylhydroperoxide (PEH) until present days in spite of the fact that ethylbenzene oxidation process was well studied and a large number of publications and books were devoted to it [2,3]. For example in the presence of homogeneous and heterogeneous catalysts on the base of transition metals compounds the selectivity of ethylbenzene oxidation into PEH is $S \leq 90\%$ at conversion degree (by spent RH) equal to $C \leq 5\%$ [3-5].

The method of homogeneous catalysts modifying by additives of electron-donor mono- and multidentate ligands for increasing of selectivity of liquid-phase oxidation processes was proposed by us.

For the first time we found the phenomenon of significant rise of initial rate (w^0), selectivity ($S = [PEH] / \Delta[RH]\cdot 100\%$) and conversion degree ($C = \Delta[RH] / [RH]^0\cdot 100\%$) of oxidation of alkylarens (ethylbenzene, cumene,) into ROOH by molecular O_2 under catalysis by transition metals complexes $M(L^1)_2$ (M = Ni(II), Co(II), L^1=acac$^-$) in the presence of additives of electron-donor monodentate ligands (L^2 = HMPA (hexamethylphosphorus triamide), dimethyl formamide (DMFA), N-methyl pyrrolidone-2 (MP)), MSt (M = Li, Na, K) [6-8]. Selectivity and conversion degree of hydrocarbon into hydroperoxide depend on hydrocarbon structure and may be regulated by selection of corresponding catalyst concentrations and ratio of system's components [8].

On the example of ethylbenzene oxidation (120°C) the mechanism of control of $M(L^1)_2$ complexes catalytic activity by additives of electron-donor monodentate ligands L^2 (L^2 = HMPA, DMFA, MP, MSt) was established.

Oxidation-reduction activity of formed in situ primary complexes $M(L^1)_2{\square}L^1$ formed in the I macro stage is increased that is expressed in the rise of rate of chain initiation (activation by O_2) and homolytic decomposition of PEH [9]. In this connection selectivity of oxidation into PEH at initial stages of reaction is not high. However initial rate of reaction is increased. With process development the increase of S_{PEH} ($S_{PEH}^{max} \approx 90\%$) in comparison with initial stage ($S_{PEH}^{max} = 80\%$) of oxidation and decrease of reaction w are observed. In developed reaction ligands L^2 control transformation of $M(L^1)_2$ complexes into more active selective particles. At that the rise of S_{PEH} is reached at the expense of catalyst participation in activation reaction of O_2 and inhibition of chain and heterolytic decomposition of PEH. Direction of side products acetophenone (AP) and methylphenylcarbinol (MPC) formation is changed from consequent (under hydroperoxide decomposition) to parallel [10, 11].

We established that in the case of nickel complexes selective catalyst was formed as result of controlled by L^2 ligand regio-selective connection of O_2 to nucleophilic carbon γ-atom of one of the ligands L^1. Coordination of electron-donor exoligand L^2 by nickel complex $Ni(L^1)_2$ (L^1=acac$^-$) promoting stabilization of intermediate zwitter-ion $L^2(L^1M(L^1)^+O_2^-)$ leads to increase of possibility of regio-selective connection of O_2 to acetylacetonate ligand activated in complex with nickel (II) ion. Further introduction of O_2 into chelate cycle accompanying by proton transfer and bonds redistribution in formed transition complex leads to break of cycle configuration with formation of (OAc^-) ion, acetaldehyde, elimination of CO and is completed by formation of catalytic particles with mixed ligands of general formula $Ni_x(acac)_y(L^1_{ox})_z(L^2)_n$ (L^1_{ox}= $MeCOO^-$) ("A") [10, 11]. Transformation of complexes $Ni(acac)_2$□L^2 (L^2=MP) leads to formation of binuclear complexes "A" with formula: $Ni_2(OAc)_3(acac)L^2$ (Scheme 1) [10]. The structure of the last ones is proved kinetically and by various physical-chemical methods of analysis (mass-spectrometry, electron and IR-spectroscopy, element analysis).

Similar change in complexes' ligand environment in consequence of acetylacetonate ligand oxidative cleavage under the action of O_2 was observed in reactions catalyzed of the only known to date a Ni(II)-containing dioxygenase – acireductone dioxygenase, ARD [12], and in reactions of oxygenation imitating the action of quercetin 2,3-dioxygenase (Cu, Fe) [13, 14].

$$L^2 \cdot L^1Ni(COMeCHMeCO)_2 + O_2 \rightarrow L^2 \cdot L^1Ni(COMeCHMeCO)^+ \ldots O_2^-$$

$$L^2 \cdot L^1Ni(COMeCHMeCO)^+ \ldots O_2^- \rightarrow L^2 \cdot L^1Ni(MeCOO) + MeCHO + CO$$

$$L^1 = (COMeCHMeCO)^-$$

$$2\ Ni(COMeCHMeCO)_2L^2 \xrightarrow{O_2} \underline{Ni_2(MeCOO)_3(COMeCHMeCO)L^2} + 3MeCHO + 3CO + L^2$$

$$L^2 = \textit{N-methylpyrrolidone-2}$$

Scheme 1.

The similarity of kinetic dependences in the parent processes of ethylbenzene oxidation in the presence of $\{Fe(III)(acac)_3 + L^2\}$ and $\{Ni(II)(acac)_2 + L^2\}$ (L^2=DMFA) (120°C) is in agreement with assumption that transformation of formed in the process $Fe(II)(acac)_2 \cdot DMFA$ complexes into more active selective catalytic species can be the result of the regioselective addition of O_2 to the γ-C atom of acetylacetonate ligand (controlled by L^2 ligand) [15]. However in this case the oxidative cleavage of the acetylacetonate ligand proceeds probably by analogy to catalysis by acetylacetone dioxygenase (M = Fe(II)) (Dke 1), which promotes dioxygen-dependent conversion of 2,4-pentandione into acetate and methylglyoxal. As in the case of catalysis by Ni complexes the active selective transformation products are hetero ligand complexes of probable structure:

$Fe(II)_x(acac)_y(OAc)_z(L^2)_n$ (L^2=DMFA) [15].

The final product of the conversion of acetylacetonate ligands is Fe(II) acetate, which as Ni(II) acetate, catalyzes heterolytic decomposition of PEH into phenol and acetaldehyde, and this is causing the S_{PEH} decrease [15].

The established mechanism of $\{Ni(L^1)_2 + L^2\}$ transformation (and probable mechanism of $\{Fe(L^1)_2 + L^2\}$ transformation) into more active selective catalysts allowed solution of the problem of selective oxidation. Methods of ethylbenzene oxidation optimization in the presence of $\{M(L^1)_2 + L^2\}$ catalytic systems were proposed and realized. Some of them are connected with increase of concentration of active mixed ligand complexes $M_x(acac)_y(L^1_{ox})_z(L^2)_n$ (M=Ni) [10] or stability of catalyst's active form, with inhibition of process of complete oxidation of "A" into inactive homogeneous-ligand complexes $M(MeCOO)_2$ carrying out heterolysis of PEH [10, 17, 18].

Problem of ethylbenzene selective oxidation into PEH can be solved using particulary the quaternary ammonium salts R_4NBr as activating exoligands-modifiers.

1. Application of Ammonium Quaternary Salts as Ligands-Modifiers for Increase of M(acac)$_n$ Activity in Reaction of Selective Oxidation of Ethylbenzene to PEH

It is evident from analysis of scheme of catalyzed oxidation including participation of catalyst in chain initiation reaction under catalyst interaction with ROOH and also in chain propagation (Cat + $RO_2^{\cdot}\rightarrow$), the chain ROOH decomposition, that with decrease of $[Cat]^0$ the rate of reaction should be decreased, and $[ROOH]^{max}$ should be increased [19, 20]. In this connection we may expect that decrease of $[M(acac)_n]^0$ will lead to increase of ethylbenzene oxidation selectivity into PEH also in the absence of L^2.

In [7, 11] we established that at low concentrations $[Ni(acac)_2]=(0.5\div1.5)\cdot10^{-4}$ mol/l (120°C) in the absence of L^2 high values of selectivity of catalyzed oxidation are possible: $S_{PEH}{}^{max}$ = 90%, but only at insignificant conversion degree, C = 4-6%. Products AP and MPC (P) are formed in this case not from PEH but parallel with PEH, i.e. $w_P / w_{PEH} \neq 0$ at $t\rightarrow0$, and furthermore

$w_{AP} / w_{MPC} \neq 0$ at $t\rightarrow0$ that indicates on parallelism of formation of AP and MPC (P = AP or MPC) [11, 18]. At these conditions addition of electron-donor monodentate ligands turned to be low effective [7] and the change of $S_{PEH}{}^{max}$ and $C_{S=90\%}$ under introduction of additives L^2 (L^2 = HMFA) into system practically was not observed.

On the base of literature data we may expect the increase of conversion degree C of catalyzed by $Ni(L^1)_2$ ($1.5\cdot10^{-4}$ mol/l) ethylbenzene oxidation (120°C) at conservation of $S_{PEH}{}^{max}$ not less than 90% in the presence of R_4NX. Coordination of R_4NX with $M(acac)_n$ may promote the oxidative transformation of $M(acac)_n$ into more active selective particles in the process of ethylbenzene oxidation into hydroperoxide by the mechanism presented in Scheme 1 (M=Ni(II)) (but in the case of Fe(II)- catalysis probably by the mechanism of catalysis by acetylacetone dioxygenase (Dke 1) (M = Fe(II)). Moreover steric factors appearing under coordination of R_4NX may hinder transformation of catalytically active hetero-ligand complex into inactive particles.

The ability of ammonium quaternary salts to complexation with transition metals compounds is also known. They proved for example that $M(acac)_2$ (M=Ni, Cu) form with R_4NX (X=$(acac)^-$, R=Me) complexes of $[R_4N][M(acac)_3]$ structure. Spectral proofs of octahedral geometry for these complexes were got [21]. Complexes Me_4NiBr_3 were synthesized and their physical properties were studied [22].

Furthermore, it is known that R_4NX in hydrocarbon mediums forms with acetylacetone complexes with strong hydrogen bond $R_4N^+(X...HOCMe{=}CHCOMe)^-$ in which acetylacetone is totally enolyzed [23]. The controlled by R_4NX regio-selective connection of O_2 by γ-C-atom of $(acac)^-$ ligand in complex $M(acac)_n\cdot R_4NX$ is probable enough. Various electrophilic reactions in complexes $R_4N^+(X...HOCMe{=}CHCOMe)^-$ proceed by γ-C-atom of acetylacetone [23, 24].

However, introduction of R_4NBr additives into ethylbenzene oxidation reaction catalyzed by complexes $Ni(L^1)_2$ ($R_4NBr = Me_4NBr$) at conditions mentioned above lead to unexpected results. Significant increase of conversion degree of oxidation into PEH at maintenance of selectivity on level S_{PEH}=90-80%, increase of S_{PEH}^{max} and initial rate of reaction w^0 were observed [25]. In the case of additives of Me_4NBr into reaction of ethylbenzene oxidation catalyzed by $Ni(II)(acac)_2$ the value of S_{PEH}^{max}=95% is higher than under catalysis by $Ni(II)(acac)_2$ without L^2. S_{PEH}^{max} is reached at C=2-3%. Selectivity remains in the limits $90\% < S \leq 95\%$ to the most deep transformation degrees of ethylbenzene C≈19% than in the presence of the other studied exoligands-modifiers for example effective multidentate ligand 18K6 (C≈12%) [18]. S_{PEH} = 95 - 80% at C=25-26%.

Additives of Me_4NBr to ethylbenzene oxidation reaction catalyzed by $Ni(L^1)_2$ lead to significant hindering of heterolysis of PEH with formation of phenol (P.) responsible for selectivity decrease.

Influence of quaternary ammonium salt on catalytic activity of $Ni(II)(acac)_2$ as selective catalyst of ethylbenzene oxidation into PEH extremely depends on radical R structure of ammonium cation. If cetyltrimethylammonium bromide (CTAB) is added S_{PEH}^{max} is reduced down to 80-82% [25]. w^0 is significantly increased, in ~ 4 times in comparison with ethylbenzene catalysis by $Ni(II)(acac)_2$ complex. Initial rate of PEH accumulation w_{PEH}^0 is higher than in the case of ehtylbenzene oxidation catalyzed by the system $\{Ni(II)(acac)_2 + Me_4NBr\}$. However initial rates of accumulation of side products of reaction of AP with MPC are also significantly increased. Selectivity decrease connected with heterolysis of PEH with phenol formation is observed at lower conversions of RH transformation.

Analysis of consequence of ethylbenzene oxidation products formation catalyzed by systems $\{Ni(II)(acac)_2+R_4NBr\}$ showed that in the course of reaction of ethylbenzene oxidation PEH, AP and MPC were formed parallel ($w_P/w_{PEH} \neq 0$ at $t \to 0$), AP and MPC were formed also parallel ($w_{AP}/w_{MPC} \neq 0$ at $t \to 0$).

Catalysis of ehtylbenzene oxidation initiated by $\{Ni(II)(acac)_2 + CTAB\}$ system is not connected with formation of micro-phase by the type of inverse micelles since the micellar effect of CTAB revealing at $t^0 < 100^0$ [26] is as a rule not important at $t^0 \geq 120^0$. Furthermore, as we saw the system $\{Ni(II)(acac)_2 + CTAB\}$ was not active in decomposition of ROOH.

The observing significant effects of S_{PEH}, C and w^0 increase are not connected with influence of Me_4NBr. We established that in the presence of one R_4NBr without nickel complex auto-catalytic developing of process with initial rates by order lower was observed, and selectivity of process by PEH equal at the beginning of reaction to 95% (Me_4NBr) was sharply reduced with the increase of ethylbenzene conversion degree. Formation of P. is observed at that from the beginning of reaction.

For estimation of catalytic activity of nickel complexes as selective catalysts of ehtylbenzene oxidation into α-phenylethylhydroperoxide we proposed to use parameter $\tilde{S}\cdot C$. $\tilde{S}$ is averaged selectivity of oxidation into PEH characterizing change of S in the course of oxidation from S^0 at the beginning of reaction to some S^{lim} conditional value chosen as a standard. For comparable by value systems selectivity as S^{lim} was selected the value S^{lim} = 80% approximately equal to selectivity of non-catalyzed ethylbenzene oxidation into PEH at initial stages of reaction. The value selectivity was carried out in the limits $S^0 \leq S \geq$

S^{lim} (S^{max} > 80%), C – conversion degree at S= S^{lim} [11, 27]. The system {Ni(II)(acac)$_2$+Me$_4$NBr} by the value of parameter $\tilde{S}\cdot C$ ($\tilde{S}\cdot C$ = 24□10^2 (%,%)) (figure 1) is the most active catalyst of ethylbenzene oxidation into PEH as compared with catalysis of studied by authors systems {Ni(II)(L^1)$_2$+L^2} [27].

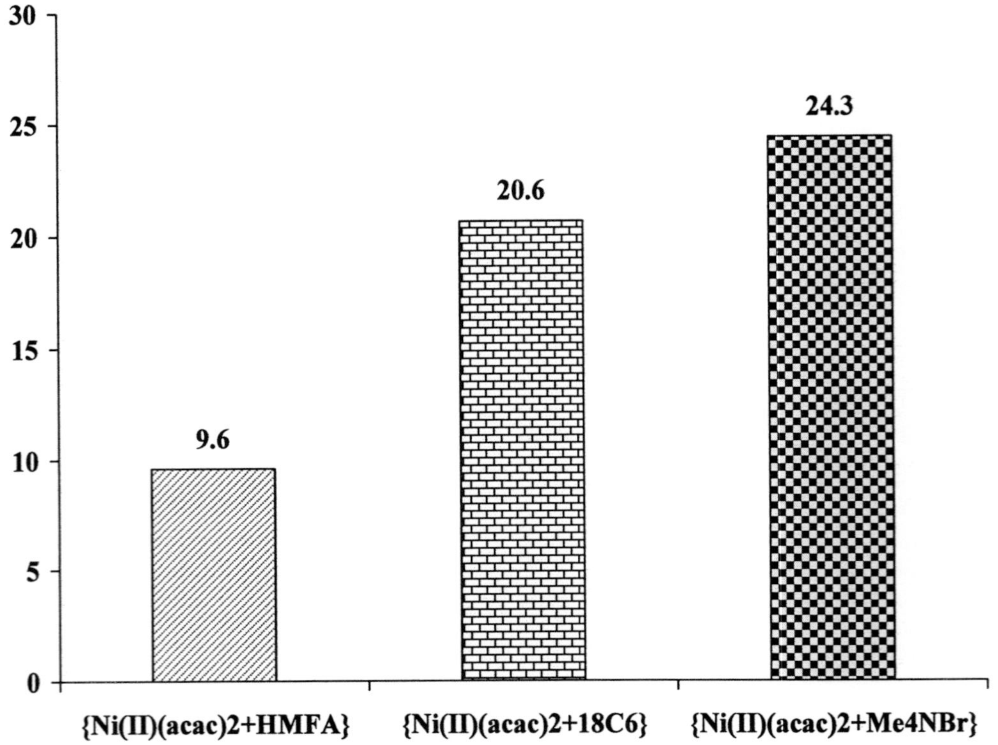

Figure 1. Parameter $\tilde{S}\cdot C$□10^{-2} (%,%) in the ethylbenzene oxidation upon catalysis by catalytic systems {Ni(II)(acac)$_2$+L^2} with L^2=Me$_4$NBr, 18C6, HMFA. [Ni(II)(acac)$_2$]=1.5·10^{-4} mol/l, 120°C.

The dependences of the initial oxidation rate w^0 in the presence of Ni(II)(acac)$_2$ on [R$_4$NBr] ([Ni(II)(acac)$_2$]=const) shows an extremum.

Synergetic effects of parameter $\tilde{S}\cdot C$ (figure 1) and w^0 (figure 2) increase indicate the formation of active complexes Ni(L^1)$_2$ with (L^2) [53] structure 1:1 (L^2=R$_4$NBr) and also the products of their transformation [25, 28].

The formation of complexes Ni(acac)$_2$ with R$_4$NBr was also proved by spectrophotometry under analysis of UV spectra of absorption of Ni(L^1)$_2$ and R$_4$NBr mixtures solutions. At that R$_4$NBr coordinate with metal ion with preservation of ligand L^1 in internal coordination sphere of complex [25]. Under formation of complexes of Ni(L^1)$_2$ with R$_4$NBr in spite of axial coordination of anion Br$^-$ by the fifth coordination place of nickel (II) ion the outer sphere coordination of ammonium cation with acetylacetonate-ion is also possible [25].

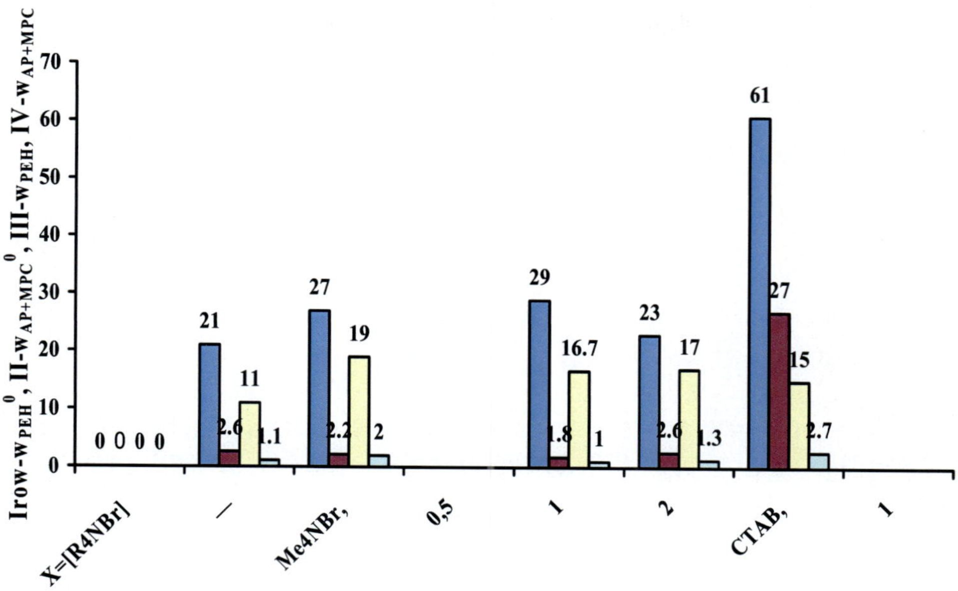

Figure 2. Initial rates w^0 (mol l^{-1} s^{-1}) and rates in the course of ethylbenzene oxidation w (mol $l^{-1}\cdot sec^{-1}$) catalyzed by {Ni(II)$(acac)_2$+R_4NBr} (L^1=$acac^{-1}$) at various concentration of [R_4NBr] (mol/l). [Ni(II)$(acac)_2$]=$1.5\cdot10^{-4}$ mol/l, 120°C.

The possibility of outer sphere coordination of R_4NX with β-diketonates of metals of variable valency was demonstrated by us using complexes Fe(III)$(acac)_3$ in the presence of various R_4NBX. In the UV spectrum Fe(III)$(acac)_3$ exhibit an intense absorption band at ν = 37 10^{-3} cm^{-1} ($CHCl_3$) of the $\pi - \pi^*$ transition of the conjugated cycle of the acetylacetonate ion [28]. In the presence of salts R_4NX (Me_4NBr, CTAB, $(C_2H_5)_4NBr$, $(C_2H_5)_3C_6H_5NCl$ and the other) a decrease in the intensity and a bathochromic shift of the absorption maximum to ν = 36 10^{-3} cm^{-1} ($\Delta\lambda \approx 10$ nm) are observed. Such a change in spectrum indicates the influence of R_4NX coordinated in the outer sphere on conjugation in the ligand. The change in the conjugation in the chelate ring of acetylacetonate complex, when R_4NX is coordinated in the outer coordination sphere of the metal can be caused by parpicipation of the oxygen atoms of the acetylacetonate ion in the formation of coordination bonds with the ammonium ion or hydrogen bonds with alkyl substituents of the ammonium ion [25].

Results of our investigation of ethylbenzene oxidation into PEH catalyzed by systems on the base of Ni(II)$(acac)_2$ and R_4NBr were used by authors of [29, 30] for developing of catalysts for industrial oxidation of ethylbenzene into PEH at the Repsol factory of Puertollano (Spain) for joint production of propylene oxide and styrene. Tetrafluorine borates of tetra-n-butylammonium and hexafluorine phosphate of 1-n-butyl-3-metylimidazolium were used as onium salts. At that S = 90-95 - 80% was observed at deep stages of oxidation not exceeding C = 10-12%, that was significantly lower than in the case of catalysis by system {Ni(II)$(acac)_2$ + Me_4NBr}. Authors [29, 30] referred on our works and used method for estimation of catalytic activity of catalysts at elementary stages of chain initiation and propagation suggested by us.

As in the case of ethylbenzene oxidation catalyzed by nickel complexes, at the catalysis by Fe(III)$(acac)_3$ the S_{PEH}^{max} increases as [Cat] is reduced. However, this increase is less significant, from S_{PEH} = 42-46% to S_{PEH} = 65%. We also observed a reduction in the rate of ethylbenzene oxidation as [Fe(III)(acac)$_3$] decreased. However, the dependence of $[PEH]^{max}$ on [Fe(III)(acac)$_3$] shows an extremum, suggesting that the mechanism of catalysis is more complicated in this case [31].

Fe(III)(acac)$_3$ and formed in the course of oxidation Fe(II)(acac)$_2$ are inactive in PEH destruction [30]. We have established, that in ethylbenzene oxidation catalyzed by Fe(III)(acac)$_3$ ([Cat]=$(0.5\div7.5)\cdot10^3$ mol/l, 80 and 120°C [31]) the major products of oxidation PEH, MPC and AP are formed parallel as at the beginning of reaction, so at deeper stages of oxidation: w_P/w_{PEH} is constant and nonzero at $t \rightarrow 0$ (P=AP+MPC) [54]. At [Cat] = $6.0\cdot10^{-3}$ and $7.5\cdot10^{-3}$ mol/l the change in the mechanism of catalysis is observed. $w_{AP}/w_{MPC} \rightarrow 0$ at $t \rightarrow 0$; in this case AP results from MPC oxidation. At [Cat] $\leq 5\cdot10^{-3}$ mol/l AP and MPC result from parallel reactions ($w_{AP}/w_{MPC} \neq 0$ at $t \rightarrow 0$). The effects of electron-donor exoligands on w, S_{PEH} and C of the ethylbenzene oxidation catalyzed by Fe(III)(acac)$_3$ were studied at [Cat] = $5\cdot10^{-3}$ mol/l. In this case [PEH] = $[PEH]^{max}$ and the kinetic changes in ethylbenzene oxidation under this conditions is not observed.

In the presence of electron-donor monodentate ligand HMPA $S_{PEH}{}^{max}$ is increased from 42 up to ~57% (80°C), conversion degree C from 5 up to 15% ([Cat] = $5\cdot10^{-3}$ mol/l), the maximum established effect of the influence electron-donor monodentate ligands on the S_{PEH} and C of ethylbenzene oxidation at catalysis by Fe(III)(acac)$_3$ [15].

In ethylbenzene oxidation reaction in the presence of {Fe(III)(acac)$_3$($5\cdot10^{-3}$ mol/l)+ R_4NBr($0.5\cdot10^{-3}$ mol/l)} (R_4NBr=CTAB) (80°C) the $S_{PEH}{}^{max}$=65% reached in the developed process is higher than in the presence of monodentate ligands HMFA, DMFA [15]. The significant growth of $[PEH]^{max}$ and increase of the w^0 rate of PEH accummulation and decrease of the w^0 rates of AP and MPC accummulation (and $[P]^{max}$) (after the autoacceleration period) are observed as compared with catalysis by Fe(III)(acac)$_3$ (figure 3A, 3B). The fast selectivity decrease of at the beginning steps of the process is connected with the transformation Fe(III) complexes in Fe(II)complexes in the course of ethylbenzene oxidation. Selectivity growth occurs at the expense of significant decrease of AP and MPC formation rate in the process obviously at parallel stages of chain propagation and chain quadratic termination (figure 3B, 4) ($w_P/w_{PEH} \neq 0$ at $t \rightarrow 0$, $w_{AP}/w_{MPC} \neq 0$ at $t \rightarrow 0$). Additives of CTAB to ethylbenzene oxidation reaction catalyzed by Fe(III)(acac)$_3$ lead to significant hindering of heterolysis of PEH with formation of phenol (P.) responsible for selectivity decrease (figure 3A). The conversion degree is increased from C = 4 up to ~ 8% (at S_{PEH} =40-65%) (figure 4).

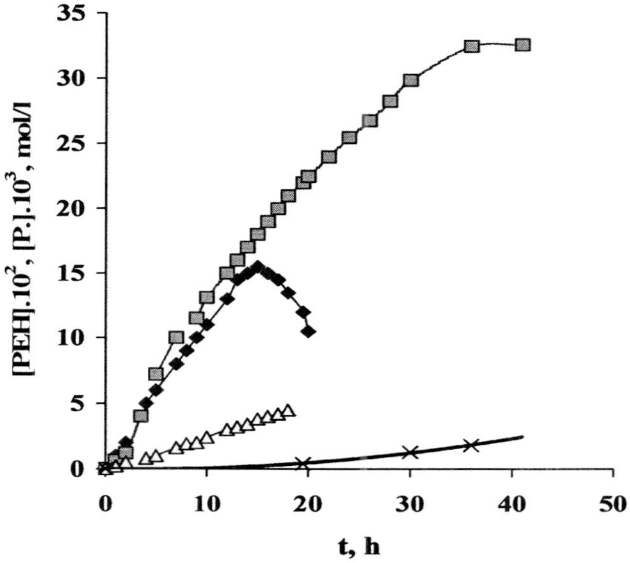

a.

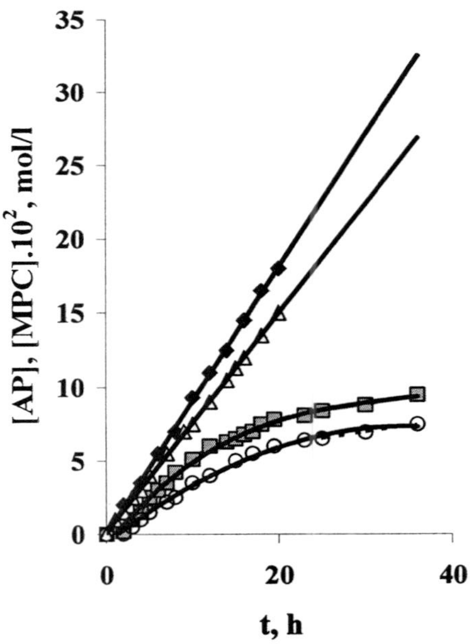

b.

Figure 3. Kinetics of accumulation of the products of ethylbenzene oxidation catalyzed by $Fe(III)(acac)_3$ and $\{Fe(III)(acac)_3$+CTAB$\}$: α-phenylethylhydroperoxide (PEH) ((♦) and (■)) and phenol (P.) ((Δ) and (x)) respectively (3 A); acetophenone (AP) ((♦) and (■)) and methylphenylcarbinol (MPC) ((Δ) and (○)) (3 B). $[Fe(III)(acac)_3]=5\cdot10^{-3}$ mol/l, $[CTAB]=0.5\cdot10^{-3}$ mol/l, 80°C.

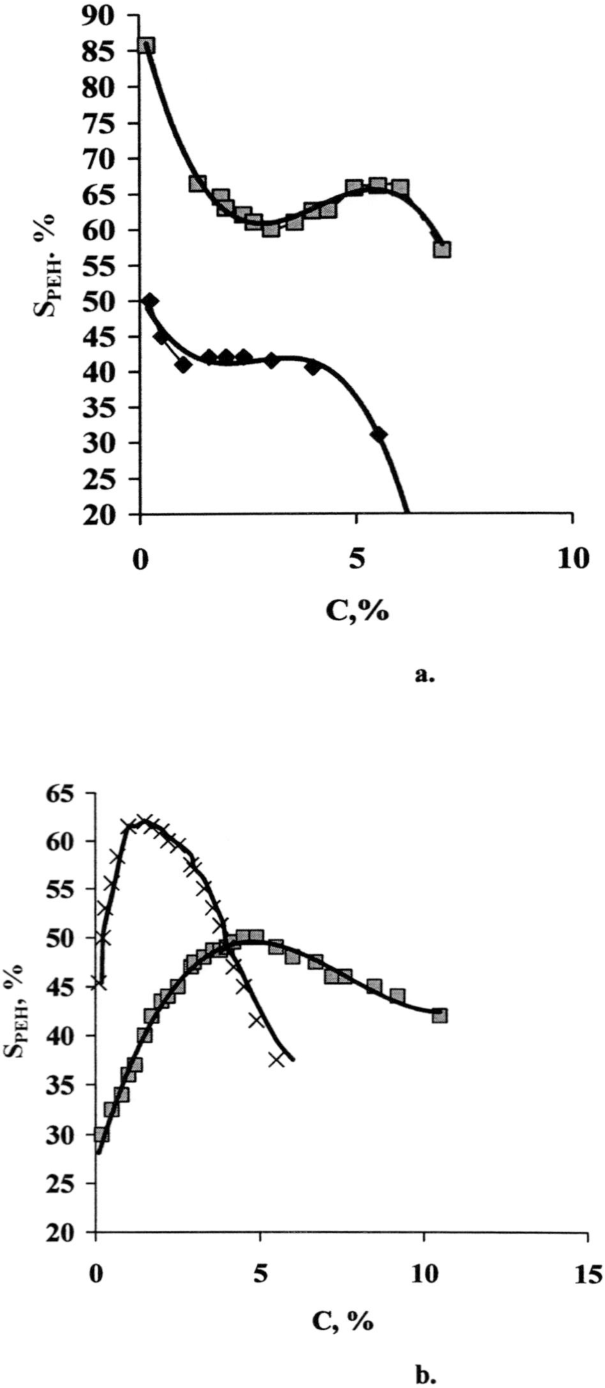

Figure 4. PEH selectivity of ethylbenzene oxidation (S_{PEH}) versus ethylbenzene conversion (C) in the presence of Fe(III)$(acac)_3$ (♦) and {Fe(III)$(acac)_3$+CTAB} (■) (4 A); {Fe(III)$(acac)_3$+Me_4NBr} (x), {Fe(III)$(acac)_3$+$(C_2H_5)_4NBr$} (■) (4 B). [Fe(III)$(acac)_3$]=$5\cdot10^{-3}$ mol/l, [R_4NBr]=$0.5\cdot10^{-3}$ mol/l, 80°C.

In the presence of {Fe(III)(acac)$_3$+CTAB} catalytic system the highest increase in the of $\tilde{S}$·C parameter is observed, in ~ 2.6, 2.36, 1.4 times for R_4NBr=CTAB, $(C_2H_5)_4NBr$), Me_4NBr) respectively in comparison with catalysis by Fe(III)(acac)$_3$ ($\tilde{S}\cdot C=2.1\cdot10^2$ (%,%)) (figure 5). In given case for value S^{lim} as standard we accept S^{lim} =40%, approximately equal to S_{PEH} for Fe(III)(acac)$_3$ in developed reaction, C – conversion degree at which $S_{PEH} \leq 40\%$ (80°C). The synergetic effects of increase in $\tilde{S}$·C parameter and w^0 observed in the reactions catalyzed by Fe(III)(acac)$_3$ in the presence of R_4NBr and also obtained kinetic regularities of ethylbenzene oxidation indicate the formation catalytic active complexes [27] presumably of $(Fe(II)(acac)_2)_x\cdot(R_4NBr)_y$ and products of their transformation in the course of ethylbenzene oxidation. Due to the favorable combination of the electronic and steric factors there is a high probability of formation of $Fe(II)_x(acac)_y(OAc)_z(CTAB)_n$ structure. Out-spherical coordination of CTAB creating sterical hindrances for regio-selective oxidation of $(acac)^-$ –ligand by the mechanism described earlier, may be more favourable in this case.

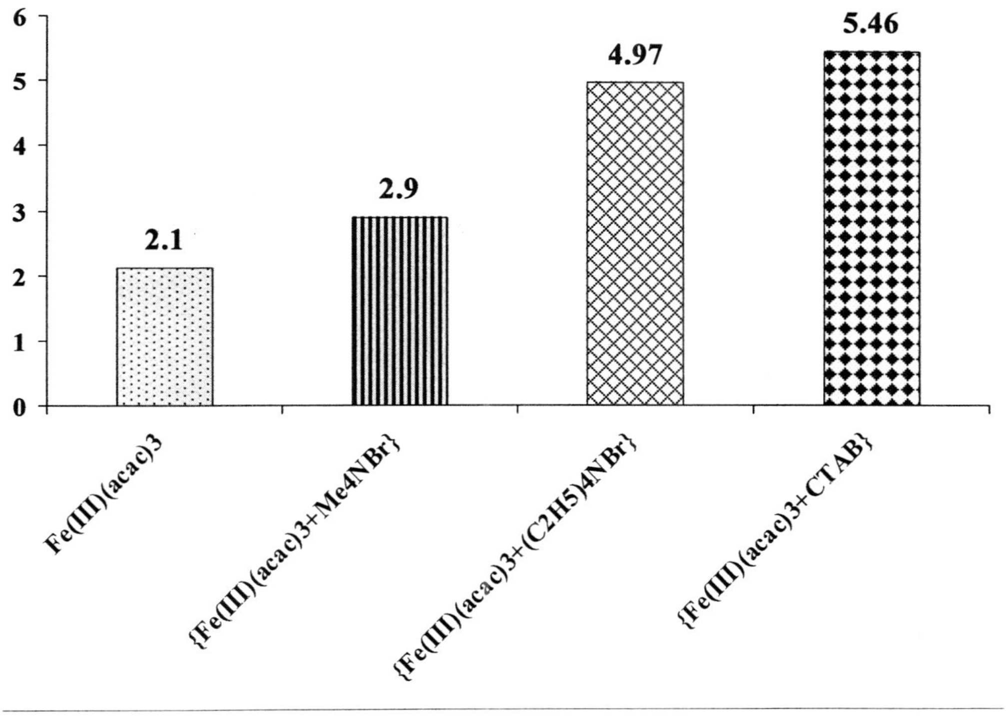

Figure 5. Parameter $\tilde{S}\cdot C\square10^{-2}$ (%,%) in the ethylbenzene oxidation upon catalysis by Fe(III)(acac)$_3$ and catalytic systems {Fe(III)(acac)$_3$+R_4NBr} with $R_4NBr=Me_4NBr$, $(C_2H_5)_4NBr$, CTAB. [Fe(III)(acac)$_3$]=$5\cdot10^{-3}$ mol/l, [R_4NBr]=$0.5\cdot10^{-3}$ mol/l, 80°C.

Catalysis of ethylbenzene oxidation initiated by {Fe(III)(acac)$_3$ + CTAB} system (80°C) in the case of application of the small concentrations R_4NBr ($0.5\cdot10^{-3}$ mol/l) evidently is not connected with formation of micro-phase by the type of inverse or sphere micelles. As we saw the system {Fe(III)(acac)$_3$ + CTAB} was not active in decomposition of PEH. $w_P/w_{PEH} \neq$

0 at t → 0, $w_{AP}/w_{MPC} \neq 0$ at t → 0. Analogous mechanism of formation PEH, AP and MPC is observed in the presence of Me_4NBr and $(C_2H_5)_4NBr$ which do not form micelles. At the [CTAB] = $5 \cdot 10^{-3}$ mol/l the rate of the PEH accumulation and $[PEH]^{max}$ decreases significantly, since the probability of micelles formation increases.

Interestingly, in the process of ethylbenzene oxidation catalyzed by {Fe(III)$(acac)_3$+$(C_2H_5)_4NBr$ ($0.5 \cdot 10^{-3}$ mol/l)} system AP is the major product of the reaction: $S_{A\Phi}$=78% at the early stages of ethylbenzene oxidation (C < 1%). [AP] » [MPC] as compared with [AP] = [MPC] in the case of catalysis by Fe(III)$(acac)_3$ without additives of L^2. At the beginning of ethylbenzene oxidation in the presence of {Fe(III)$(acac)_3$+$(C_2H_5)_4NBr$ ($0.5 \cdot 10^{-3}$ mol/l)} (C ≈ 2%) [AP]/ [MPC] > 10, [MPC]/ [PEH] ≈ 3.6 (figure 3 A, 3 B). In the presence of {Fe(III)$(acac)_3$+Me_4NBr ($0.5 \cdot 10^{-3}$ mol/l)} (C ≤ 2%)) [AP] > [MPC] also. But [AP]/ [MPC] ≈ 4-5, and S_{AP}=36% (C < 1%) ~ as in the case of catalysis by Fe(III)$(acac)_3$ (S_{AP}=30-33% (C = 1-2%)).

Thus, we established the interesting fact – the catalytic effect of small concentrations of quaternary ammonium salts, $[R_4NBr]$ = $0.5 \cdot 10^{-3}$ mol/l, which in 10 times less than [Fe(III)$(acac)_3$]. It is known that salts QX can form complexes with metal compounds of variable composition which depends on the nature of solvent [32].

2. Participation of Catalysts Active Forms in Elementary Stages of Radical-Chain Ethylbenzene Oxidation Catalyzed by {$M(L^1)_2$+R_4NBr} (M=Ni(II), Fe(III), L^1=$acac^{-1}$)

We propose the method for estimation of catalytic activity of complexes formed *in situ* at the beginning of reaction and in developed process, at elementary stages of oxidation process [11, 15, 18, 25] by simplified scheme assuming quadratic termination of chain and equality to zero of rate of homolytic decomposition of ROOH. In the framework of radical-chain mechanism the chain termination rate in this case will be (1):

$$w_{term}=k_6[RO_2^\bullet]^2=k_6\left\{\frac{\mathbf{w}_{PEH}}{\mathbf{k}_2[RH]}\right\}^2 \qquad (1)$$

where w_{PEH} – rate of PEH accumulation, k_6 – constant of reaction rate of quadratic chain termination; k_2 – constant of rate of chain propagation reaction $RO_2^\bullet$ + RH→.

We established that complexes $M(L^1)_n$ (M=Ni(II) ([Cat]=$(0.5\text{-}1.5) \cdot 10^{-4}$ mol/l), Fe(III) ([Cat]=$(0.5\text{-}5) \cdot 10^{-3}$ mol/l)) were inactive in PEH homolysis, products MPC and AP were formed at stages of chain propagation Cat + $RO_2^\bullet$→ and quadratic termination of chain. Actually, $w^0 \sim [Cat]^{1/2}$ and $w_i^0 \sim [Cat]$ and linear radicals termination on catalyst may be not taken into account [20]. In the case of quasi-stationarity by radicals $RO_2^\bullet$ the values $w_{term.}=w_i$ are the measures of nickel(II) and iron(II) complexes activity in relation to molecular O_2. Discrepancy between w_{AP+MPC} and w_{term} (figure 6, table) in the case of absence of linear termination of chain is connected with additional formation of alcohol and ketone at the stage of chain propagation Cat + $RO_2^{\bullet} \rightarrow$ (2):

$$w_{pr.} = w_{AP+MPC} - w_{term} \quad (2)$$

The direct proportional dependence of w_{pr}^0 on [Cat] testifies in favour of nickel(II) and iron(II) complexes participation at stage of chain propagation Cat + $RO_2^{\bullet} \rightarrow$.

Table. Rates of accumulation of ethylbenzene oxidation products (mol l^{-1} s^{-1}) at the beginning of reaction (w^0), and calculated rates of chain initiation (w_i^0) and chain propagation (w_{pr}^0) and ($w_i^0/w_{пp}^0$)·100%. Catalysis by Fe(III)(acac)$_3$ and {Fe(III)(acac)$_3$+R$_4$NBr}. [Fe(III)(acac)$_3$] = 5·10^{-3} mol/l, [R$_4$NBr] = 0.5·10^{-3} mol/l. 80^0 C.

Cat	$w_{PEH}^0 \cdot 10^6$	$w_{AP+MPC}^0\ 10^6$	$w_i^0 \cdot 10^7$	$w_{pr}^0 \cdot 10^6$	(w_i^0/w_{pr}^0) 100%
Fe(III)(acac)$_3$	2.90	3.40	0.79	3.32	2.38
{Fe(III)(acac)$_3$+(C$_2$H$_5$)$_4$NBr}	5.90	8.30	3.0	8.0	3.75
{Fe(III)(acac)$_3$+ Me$_4$NBr}	4.20	3.40	1.52	3.25	4.67
{Fe(III)(acac)$_3$+CTAB}	4.35	3.30	1.63	3.14	5.19

*$w_P = w_{AP+MPC}$

We suggest that these conditions $w^0 \sim [Cat]^{1/2}$ and $w_i^0 \sim [Cat]$ will be fulfilled also in the presence of additives of R$_4$NBr.

Except theoretical consideration in [3], activity of transition metals complexes $M(L^1)_n$ (M=Ni, Co, Fe, L^1 = acac$^-$, enamac$^-$) at stage of chain propagation (Cat + $RO_2^{\bullet} \rightarrow$) of ethylbenzene oxidation is estimated only in our works [11, 15, 18, 25, 31].

Investigation of reaction ability of peroxide complexes [LM-OOR] (M=Co, Fe) preliminary synthesized by reactions of compounds of Co and Fe with ROOH or $RO_2^{\bullet}$-radicals [33-35] confirms their participation as intermediates in reactions of hydrocarbons oxidation. Obviously, the schemes of radical-chain oxidation including intermediate formation [LM-OOR] [33-36] with further homolytic decomposition of peroxo-complexes ([LM-OOR]→R'C=O (ROH) + $R^{\bullet}$) may explain parallel formation of alcohol and ketone under ethylbenzene oxidation in the presence of $M(L^1)_n$ (L^1 = acac$^-$) and their complexes with R$_4$NBr.

We established that mechanism of selective catalysis of complexes $M(L^1)_2$ (M=Ni, Fe) and products of their transformation depended on both ratio of rates of chain initiation w_i (activation by O$_2$) and propagation (w_{pr}) and on activity of Cat in PEH decomposition (homolysis, heterolysis of PEH, chain decomposition of PEH).

In non-catalytic ethylbenzene oxidation at high temperatures the formation of active sites occurs in reaction of chain initiation (RH+O$_2$→) and under chain decomposition of PEH, the value S_{PEH} to a significant extent should be determined by factor of instability of PEH $\beta = w_{PEH}^- / w_{PEH}^+$ (w_{PEH}^- – sum rate of PEH decomposition (thermal (molecular) and chain), w_{PEH}^+ – rate of chain PEH formation). Actually, it turned out that value β in the course of non-catalyzed process of ethylbenzene oxidation is increased at the expense of rise of PEH chain decomposition rate that leads to reduction of S_{PEH} [2, 3].

At conditions of ethylbenzene oxidation catalyzed by $M(acac)_2$ (M=Ni(II), Fe(II)) and $M(acac)_n$ complexes with R_4NBr the value is $\beta = w_{PEH}^-/w_{PEH}^+ \rightarrow 0$ as at the beginning, so in developed process, the direction of AP and MPC formation is changed (consequent (from PEH decomposition) → parallel), S_{PEH} depends on catalyst activity at stages of chain initiation (activation by O_2) and propagation Cat+ $RO_2^\bullet \rightarrow$.

Calculation by formulas (1) and (2) show that high activity of "primary" complexes $Ni(II)(acac)_2 \cdot R_4NBr$ as selective catalysts of ethylbenzene into PEH oxidation is connected with growth of rate w_i^0 in comparison with catalysis by $Ni(L^I)_2$, hindering of rate of chain propagation (w_{pr}^0) Cat + $RO_2^\bullet \rightarrow$ (figure 6) (and also homolysis and heterolysis of PEH).

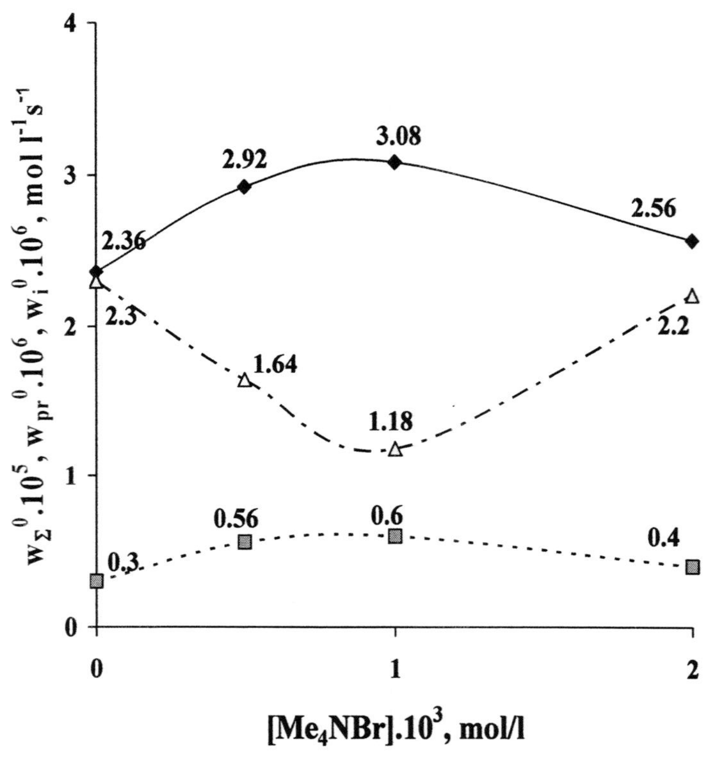

Figure 6. The ethylbenzene oxidation rates at the beginning of reaction $\mathbf{w_\Sigma^0}$ (♦), and calculated rates of chain initiation $\mathbf{w_i^0}$ (■), and propagation $\mathbf{w_{np}^0}$ (Δ) as a function of $[Me_4NBr]$ in the ethylbenzene oxidation upon catalysis by {$Ni(II)(acac)_2$+Me_4NBr} $[Ni(II)(acac)_2] = 1.5 \cdot 10^{-4}$ mol/l, 120^0 C.

It can be seen from figure 6 that the minimum value of w_{pr}^0 is observed at $[Me_4NBr] = 1 \cdot 10^{-3}$ mol/l, with corresponds to the formation of the complex in 1:1 ratio. As $[Me_4NBr]$ increases further w_{pr}^0 increases too. At $[Me_4NBr] = 2 \cdot 10^{-3}$ mol/l the latter reaches the value of w_{pr}^0 observed in the presence of $Ni(II)(acac)_2$ only. It is evident that in the process of Me_4NBr coordination (in the inner and in outer spheres) steric hindrances to coordination of $RO_2^\bullet$ with the metal ion can appear (w_{pr}^0 drops). If $[Me_4NBr]$ is rather large, the probability of opening of the chelate ring of the $(acac)^-$ ion increases and coordination of the radical $RO_2^\cdot$ with the metal center becomes possible (w_{pr}^0 increases).

Under substitution of radical CH_3 (Me_4NBr) in cation R_4N^+ by radical n-$C_{16}H_{33}$ (CTAB) the value of S_{PEH}^0 is reduced (from S_{PEH}^0=92% up to S_{PEH}^0=70%). Activity of formed complexes Ni(II)$(acac)_2$·CTAB at stages of chain initiation and propagation is increased in 4,6 and in 20,5 times correspondingly. At that the rate of PEH accumulation (w^0_{PEH}) is increased only in 2 times, and w^0_{AP+MPC} in 15.4 times in comparison with catalysis by Ni(II)$(acac)_2$·Me_4NBr (figure 2).

Higher value of $\tilde{S}$·C parameter in the case of ethylbenzene oxidation catalysis by systems {Ni(II)$(acac)_2$+Me_4NBr} in comparison with {Ni(II)$(acac)_2$+L^2} (L^2=HMFA, 18K6) (figure 1) may be explained by higher stability of formed in the course of reaction catalytic particles, presumably $Ni_x(acac)_y(OAc)_z$·Me_4NBr to complete oxidation to nickel acetate. Out-spherical coordination of Me_4NBr creating sterical hindrances for regio-selective oxidation of $(acac)^-$ –ligand may be favourable to the last fact [25].

Maintenance of high selectivity of developed reaction catalyzed by systems {Ni(II)$(L^1)_2$+R_4NBr} in comparison with non catalyzed oxidation of ethylbenzene (S_{PEH} << 80%) is connected mainly with hindering of chain and heterolytic decomposition of PEH and change of AP, MPC products formation mechanism (consequent→parallel).

The chain initiation in ethylbenzene oxidation by dioxygen in the presence of Fe(III)$(acac)_3$ and {Fe(III)$(acac)_3$+ R_4NBr} can be represented by the following reaction:

$$\mathrm{Fe(III)(acac)_3\ ((Fe(III)(acac)_2)_m\cdot(R_4NBr)_n) + RH \rightarrow}$$
$$\mathrm{\rightarrow Fe(II)(acac)_2\ ((Fe(II)(acac)_2)_x\cdot(R_4NBr)_y)\ \ldots\ Hacac + R^{\bullet}} \qquad \text{(I)}$$

The reaction Fe(III) and interaction of the resulting Fe(II) complex with dioxygen appear to be responsible for chain initiation in the reaction catalyzed by Fe(III)$(acac)_3$ and {Fe(III)$(acac)_3$+ R_4NBr}.

As is seen from the table, the rate of chain initiation in the presence of {Fe(III)$(acac)_3$+ R_4NBr} is higher than in the reaction catalyzed by Fe(III)$(acac)_3$ and much higher than in the noncatalytic reaction ($w_i^0 \approx 10^{-9}$ mol l^{-1} s^{-1}).

The ratio $w_i^0/\ w_{pr}^0 \approx$ 2-5% ({Fe(III)$(acac)_3$+ L^2} L^2=R_4NBr, HMFA, DMFA [15]) means that the iron complexes are more active in chain propagation (Cat + $RO_2^{\bullet}$→) than in chain initiation. Furthermore, this value indicates that the reaction Cat + $RO_2^{\bullet}$→ plays a greater role in ethylbenzene oxidation catalyzed by the iron complexes than in the same reaction catalyzed by the Ni(II) complexes. In the latter case, the ratio $w_i^0/\ w_{pr}^0 \approx$ 11-50%, and depends on the nature of the ligand environment of metal ion [11, 18, 25]. The estimated ratio $w_i^0/\ w_{pr}^0 \approx$ 2-5% suggests that AP and MPC form mainly in the chain propagation step (conceivably, through the homolytic decomposition of the intermediate complex [$L^2Fe(L^1)_2$—OOR] [33-36]). Furthermore, this ratio points to an essential contribution from the chain propagation to the overall rate of the process. The structure of radical R in the R_4N^+ cation evidently has influence on the catalyst activity at stages of chain initiation (activation by O_2) and propagation Cat + $RO_2^{\bullet}$→ (table).

S_{PEH}^0 depends on the catalyst activity at stages of chain initiation (activation by O_2) and propagation Cat + $RO_2^{\bullet}$→ at catalysis by Fe(III)$(acac)_3$ in the presence of R_4NBr. Thus, the rise of S_{PEH}^0 from 45 up to ~ 65% in ethylbenzene oxidation catalyzed by {Fe(III)$(acac)_3$+R_4NBr} (R_4NBr=CTAB) system (table) is connected with an increase in w_i^0

(in 2 times) and a small decrease in w_{pr}^0. The rate of PEH formation increases in ~1.5 times and the rate AP and MPC formation at stages of chain propagation Cat + $RO_2^{\bullet}\rightarrow$ increases insignificant. The analogous effects are observed with R_4NBr=Me_4NBr. In the case of $(C_2H_5)_4NBr$ the decrease in the S_{PEH}^0 is due to the growth of the rates w_{AP+MPC}^0 (w_{pr}^0) as compared with an increase in the rates w_{PEH}^0. The complexes $Fe(II)_x(acac)_y(OAc)_z(R_4NBr)_n$ formed in the process of ethylbenzene oxidation catalyzed by system {$Fe(III)(acac)_3$+R_4NBr} are likely to be inactive in the the heterolytic decomposition of PEH and in the reaction with the $RO_2^{\bullet}$ radicals (figure 3, 4).

Thus the established by us mechanism of catalysis by nickel complexes $Ni(II)(L^1)_2$ (L^1=$acac^-$) in the presence of additives of electron-donor monodentate ligands-modifiers L^2 (L^2=HMFA, DMFA, MP, MSt) in the ethylbenzene oxidation by dioxygen allowed proposing the methods for control of selective oxidation process, in particulary, the use of additives of quaternary ammonium salts R_4NBr as activating exoligands-modifiers. Values of selectivity, S_{PEH}, conversion degree, C, and PEH yield reached at that exceed analogous parameters in the presence of the other {$Ni(II)(L^1)_2$+L^2} systems studied by us and known catalysts of ethylbenzene oxidation into PEH. Mechanism of activation of catalytic system {$Ni(II)(L^1)_2$+R_4NBr} consists in increase of activity of primary complexes $Ni(II)(L^1)_2\cdot R_4NBr$ as selective catalysts; stabilization of active selective complexes $Ni(II)_x(L^1)_y(L^1{}_{ox})_z\cdot R_4NBr$, the products of $Ni(II)(L^1)_2\cdot R_4NBr$ oxygenation, formed in the course of oxidation.

The relatively low efficiency of $Fe(III)(acac)_3$ and {$Fe(III)(acac)_3$+L^2} systems (as a selective catalysts for ethylbenzene oxidation to PEH) as compared to the efficiency of Ni(II) complexes is due to the simultaneous formation of PEH, MPC, and AP. The principle possibility of the composition catalysts construction on the basis of complexes $Fe(III)(acac)_3$ and R_4NBr as ligands-modifiers for the ethylbenzene oxidation selectivity increase by dioxygen into PEH is demonstrated.

The value method of activity of the constructed catalytic systems on the basis of nickel and ferrous complexes in the micro steps of the chain-radical ethylbenzene oxidation is offered. As was established the rate and selectivity of ethylbenzene oxidation into PEH at the catalysis by initial complexes depended on the ratio between rates of chain initiation (w_i, activation O_2) and propagation (w_{pr}, Cat+$RO_2^{\bullet}\rightarrow$), w_i / w_{pr}. The selective catalysis by catalytic particles formed in the course of oxidation process depended also on activity of Cat in PEH decomposition (chain, homolytic, heterolytic), direction of products formation methylphenylcarbinole and acetophenone (parallel with PEH or consecutively under hydroperoxide decomposition).

The ratio $w_i^0/ w_{pr}^0 \approx$ 2-5% ({$Fe(III)(acac)_3$+ L^2} L^2=R_4NBr, HMFA,DMFA [15]) means that the iron complexes are more active in chain propagation (Cat + $RO_2^{\bullet}\rightarrow$) than in chain initiation. Furthermore, this value indicates that the reaction Cat + $RO_2^{\bullet}\rightarrow$ plays a greater role in ethylbenzene oxidation catalyzed by the iron complexes than in the same reaction catalyzed by the nickel complexes. In the latter case, the ratio $w_i^0/ w_{pr}^0 \approx$ 11-50%, and depends on the nature of the ligand environment of metal ion. The rise of S_{PEH}^0 from 45 up to ~ 65% in ethylbenzene oxidation catalyzed by {$Fe(III)(acac)_3$+R_4NBr} (R_4NBr=CTAB) system (table) is connected with increasing

w_i^0 (w_i^0/ w_{pr}^0).

The 'dioxigenase-like' mechanism of the transformation of the $(Fe(II)(L^1)_2)_x(R_4NBr)_y$ complexes formed at the early stages of the ethylbenzene oxidation at catalysis by $\{Fe(III)(acac)_3+R_4NBr\}$ systems into new catalysts in the course of oxidation process is proposed.

REFERENCES

[1] K. Weissermel, H.-J. Arpe, *Industrial Organic Chemistry,* 3nd ed., transl. by C.R. Lindley. New York, VCH, 1997, 427 pp.

[2] N.M. Emanuel, *Uspekhi khimii,* Vol. 47, N 8, pp. 1329-1396, 1978 (in Russian).

[3] N.M. Emanuel, D.Gal, Ethylbenzene oxidation. Model reaction, Moscow, *Nauka,* 1984, 375 pp. (in Russian).

[4] Yu.D. Norikov, E.A. Blyumberg, L.V. Salukvadze, Problemy kinetiki i kataliza, Moscow, *Nauka,* Vol. 16, pp.150-165, 1975 (in Russian).

[5] M.V. Nesterov, V.A. Ivanov, V.M. Potekhin, V.A. Proskuryakov, M.Yu. Lysukhin, *Zh. Prikl. Khimii,* Vol. 52, N 7, pp. 1585-1589, 1979 (in Russian).

[6] L.A. Mosolova, L.I. Matienko, Z.K. Maizus, *Izv. AN SSSR, Ser. Khim.,* N 2, pp. 278-281, 1980 (in Russian).

[7] L.A. Mosolova, L.I. Matienko, *Neftekhimiya,* Vol. 25, N 4, pp. 540-545, 1985 (in Russian).

[8] L.A. Mosolova, L.I. Matienko, I.P. Skibida, *Kinetika i kataliz,* Vol. 29, N 5, pp. 1078-1083, 1988 (in Russian).

[9] L.A. Mosolova, L.I. Matienko, Z.K. Maizus, *Izv. AN SSSR, Ser. Khim.,* N 4, pp. 731-735, 1981 (in Russian).

[10] L.A. Mosolova, L.I. Matienko, I.P. Skibida, *Kinetika i kataliz,* Vol. 28, N 2, pp. 479-484, 1987 (in Russian).

[11] L.I. Matienko, L.A. Mosolova, Izv. *AN, Ser. Khim.,* N 1, pp.55-60, 1999 (in Russian).

[12] Y. Dai, Th. C. Pochapsky, R.H. Abeles, *Biochem.,* Vol. 40. N 21. pp. 6379-6387, 2001.

[13] M. Utaka, M. Hojo, Y. Fujiu, A. Takeda, *Chem. Lett.,* N 6, pp. 635-638, 1991.

[14] T. Sagawa, K. Ohkibo, *J. Mol. Catal. A: Chem.,* Vol. 113, pp. 269-281, 1996.

[15] L.I. Matienko, L.A. Mosolova, *Neftekhimiya* (in prints).

[16] G. D. Straganz, B. Nidetzky, *J. Am. Chem. Soc., Vol.* 127. N 35. pp. 12306-12314, 2005.

[17] L.A. Mosolova, L.I. Matienko, I.P. Skibida, *Kinetika i kataliz,* Vol. 28, N 2, pp. 484-487, 1987 (in Russian).

[18] L.A. Mosolova, L.I. Matienko, I.P. Skibida, *Izv. AN, Ser. Khim.,* N 8, pp.1406-1410, 1994 (in Russian).

[19] E.T. Denisov, N.M. Emanuel, *Uspekhi khimii, Vol.* 29, N 12, pp. 1409-1438, 1960 (in Russian).

[20] N.M. Emanuel, E.T. Denisov, Z.K. Maizus, Chain reactions of hydrocarbons oxidation in liquid phase, Moscow, *Nauka,* 1965, 375pp. (in Russian).

[21] Ms.M. Satpathy, B. Pradhan, *Asian J. Chem.,* Vol. 9, N 4, pp. 873-876, 1997.

[22] H. Yamaguchi., K. Katsumata, M. Steiner, H.J. Miketa, *J. Magn. Magn. Mater., Vol.* 177-181 (Pt.1), pp. 750-751, 1998.

[23] J.H. Clark, J.M. Miller, *J. Chem. Soc. Perkin Trans.*, N 1, pp. 1743-1750, 1977.

[24] E.V. Demlov, *Izv. AN, Ser. Khim.*, N 11, pp. 2094-2101, 1995 (in Russian).

[25] L.A. Mosolova, L.I. Matienko, I.P. Skibida, *Izv. AN, Ser. Khim.*, N 8, pp.1412-1416, 1994 (in Russian).

[26]. T.V. Maximova, T.V. Sirota, E.V. Koverzanova, O.T. Kasaikina, *Neftekhimiya,* Vol. 41, N 4, pp. 289-293, 2001 (in Russian).

[27] V.A. Golodov, *Ross. Khim. Zh., Vol.* 44, N 3, pp 45-57, 2000 (in Russian).

[28] R.L.Litvendt, L.K.Kernitsky, Ino*rg. Chem., Vol.* 9, N 3, pp. 491-494, 1970.

[29] R. Alcantara, L. Canoira, P. Guilherme-Joao, J.-M. Santos, I. Vazquez, *Appl. Catal. A: Gener., Vol.* 203, pp. 259-268, 2000.

[30] R. Alcantara, L. Canoira, P. Guilherme-Joao, P. Perez-Mendo, *Appl. Catal. A: Gener., Vol.* 218, pp. 269-279, 2001.

[31] L.I. Matienko, L.A. Mosolova, *Kinetika i kataliz, Vol.* 46, N 3, pp. 354-360, 2005 (in Russian).

[32] Z.G. Aliev, A.M. Karateev, L.O. Atovmjian, *Koord. Chim., Vol.* 14, N 4, pp. 530-535, 1988 (in Russian).

[33] R.D.Arasasingham, Ch.R. Cornman, A.L. Balch, *J. Am. Chem. Soc., Vol.* 111, N 20, pp. 7800-7805, 1989.

[34] F.A. Chaves, J.M. Rowland, M.M. Olmstead, P.K. Mascharak, *J. Am. Chem. Soc., Vol.* 120, N 35, pp. 9015-9027, 1998.

[35] E. Solomon-Rapaport, A. Masarwa, H. Cohen, D. Meyerstein, *Inorg. Chim. Acta.,* N 299, pp. 41-46, 2000.

[36] A.E. Semenchenko, V.M. Solyanikov, E.T. Denisov, *Zh. Phiz. Khimii, Vol.* 47, N 5, pp. 1148-1151, 1973 (in Russian).

In: New Aspects of Biochemical Physics
Editors: Sergei D. Varfolomeev et al., pp. 113-121
ISBN 1-60021-419-3

Chapter 16

INFLUENCE OF LOW-TOXICITY CHEMICAL AGENTS AT LOW DOSES ON OXIDATIVE PROCESSES IN THE LIVER OF MICE

M.V. Kozlov, V.V. Urnysheva and L.N. Shishkina
N.M. Emanuel Institute of Biochemical Physics, Russian Academy of Sciences,
ul. Kosygin 4, Moscow, 119991 Russia

ABSTRACT

It was shown that the scale or direction of correlations between both the several parameters of the lipid composition and the generalized parameters of the phospholipids composition of the murine liver depend on the kinetic characteristics of their lipids (the lipid peroxides content, the capacity of lipids to decompose peroxides). In was found that the combined intraperitoneal administration of the low- toxicity surfactant (Tween 80) and acetone at low doses causes disturbances in the lipid peroxidation regulatory system which manifest themselves in 1 month after the action.

Key words: lipid peroxidation, low doses, liver of mice, Tween 80, acetone, combined effect

AIMS AND BACKGROUND

The hematopoietic system and hepatocytes of liver play a important role in the maintenance of the homeostasis in organism. In [1, 2] it was shown that hepatocytes are also having high sensitive to technogenic contamination of the environment. The detoxification which consists in the removal of various chemical substances that are products of the vitally important processes in organism and alien compounds is one of the main function of liver [3].

The investigation of the state of lipid peroxidation (LPO) is one of the perspective lead in the search for tests for the estimation of biological consequences of damaging effects because of a high sensitivity of the LPO regulatory system parameters in tissues of wild rodents and laboratory animals to the action of unfourable factors [1, 2, 4, 5].

Polyoxyethylensorbitanmonooleate (Tween 80) which is one of the common technogenic unfourable factors – surfactant and acetone were chosen as chemical agents. The surfactant's role in the development of biological consequences of the combined effect of different agents enhances due to their properties to facilitate the penetration of hydrophobic compounds into cells. In practice, this distinction of surfactants is used for the prophylactic or therapeutic administration of lipid soluble substances into organism [6, 7]. Acetone is also one of the chemical agents which is widely used as a solvent and intermediate in the chemical, pharmaceutic and food industry.

It is well known that liver is the main organ of accumulation and metabolism of chemical agents in organism. This organ was chosen as an object for investigating of the remote consequences of the action of the chemical agents on organism.

The aim of this work is to study the combined effect of the surfactant and acetone at low doses on the parameters of the LPO regulatory system in the murine liver.

EXPERIMENTAL

The experiments were carried out on Balb/c mice (males, weight 16 – 20 g before the experiments, n = 13). The mice were divided into two groups. The first group (8 mice) was used as the intact biological control. The Tween 80 (Ferak Berlin, FRG) at the dose of 30 mg/kg as 0.3% solution in 10% mixture of acetone in water was administration intraperitoneally at the second group of mice (5 mice). The solution of 0.3% Tween 80 in the 10% aqueous solution of acetone (special purity grade) was prepared immediately before the experiment. Decapitation of mice was performed in 1 month after administration of chemical agents. Decapitation of intact mice was performed simultaneously with the experimental group. The murine liver was placed in ice-cooled weighting bottles immediately after the decapitation. The parameters to be studied were determined for each mouse.

Lipids of the liver were isolated by the method of Blay and Dyer in the Kates modification [8]. The qualitative and quantitative composition of phospholipids (PL) was determined by thin-layer chromatography as it was described in [9]. We used type G silica gel (Sigma, USA), glass plates 9 × 12 cm and the mixture of solvents - chloroform: methanol: glacial acetic acid: water (25:15:4:1) as a mobile phase. All the solvents were of special or chemical purity grade. The development of chromatograms was performed by iodine vapour. For the colour reaction for phosphorus, we used ammonium molybdate and ascorbic acid (Serva, FRG) and chemical purity perchloric acid. The amount of inorganic phosphorus was determined from the optic density of the solutions at the wavelength $\lambda = 800$ nm with KFK-3 (Russia) instrument. In addition to the quantitative analysis of different fractions of PL, the following generalized parameters of the lipid composition were also evaluated: the PL proportion in the total lipid composition (% PL); the ratio of phosphatidyl choline to phosphatidyl ethanolamine (PC/PE) content in PL; the ratio of the sums of the more easily to the more poorly oxidizable fractions of PL (ΣEOPLΣPOPL). The ΣEOPLΣPOPL value was calculated by the formula: ΣEOPLΣPOPL = (PI + PS + PE + CL + PA)/(LPC + SM + PC), where PI is phosphatidyl inositol, PS is phosphatidyl serine, CL is cardiolipin, PA is phosphatidic acid, LPC are PL lysoforms and SM is sphingomielin.

The content of LPO products, which have reacted with 2-thiobarbituric acid (TBA-reactive substances, TBA-RS) was determined spectrophotometrically by the method described in [10] with adding 10 μl of the 0.01% 4-methyl-2,6–ditert.butylphenol solution in ethanol. Protein was determined according to [11]. The peroxides concentration in lipids was determined iodometrically.

The data were processed by a commonly used variational statistic method [12] and by the KINS program given in [13].

Results and Discussion

One of the primary significance in the development of both the early and remote biological consequences of those effects have changes in the hematopoietic system induced by damaging effects. The inevitable changes in the organ weight observed under different experimental conditions made it possible to use these parameters as a criterion for the organ activity [14]. In larger animals, the organ weight is larger. As expected there is a direct correlation between the weight of liver and that of a mouse for the age control group with the correlation coefficient $R = 0.91 \pm 0.06$. However, in 1 month after the administration of surfactant and acetone the correlation between these parameters is absent for this group of mice ($R = 0.32$). This is evidence for a long-term persistence of the disturbances in the liver activity on the physiological level after the combined effect of chemical agents at the low doses. It should be noted that the variability of the liver weight in the experimental group decreased drastically as compared with that for the control group. Visually the murine liver in the experimental group was dark and the size of the gallbladder is smaller than in the control group.

The influence of the unfourable low-toxicity ecological factors of the chemical nature at low doses on the LPO process in Balb/c mice liver was assessed.

From our experiments we found that the liver lipids of mice in the control age group exhibit different kinetic characteristics. Liver lipids of one subgroup of the intact animals contain peroxides ($[ROOH]_0$), liver lipids of another control subgroup had a capacity of decomposing peroxides, i.e. exhibit an antiperoxide activity (APA). In these cases the concentration of peroxide in liver lipids is equal to zero. It was found that depending on the kinetic characteristics of lipids there are differences in values of the several generalized parameters of the lipid composition and the scale and direction of correlation between these parameters in these groups of control mice. Thus, when lipids exhibit the property of the peroxide decomposition, the average values of the PL fraction in the total lipid content and the ratio of the sums of the more easily to the more poorly oxidizable PL fractions are significantly lower as compared with the control animals whose lipids contain peroxides (figure 1). The average value of the PC/PE ratio in the group of mice with APA of their liver lipids also tends to decrease as compared with that for the another control animals. On the contrary, the concentration of the oxidation products ([TBA-RS]) in the subgroup of mice with APA of their liver lipids tends to increase as compared with that for the another control animals (figure 1).

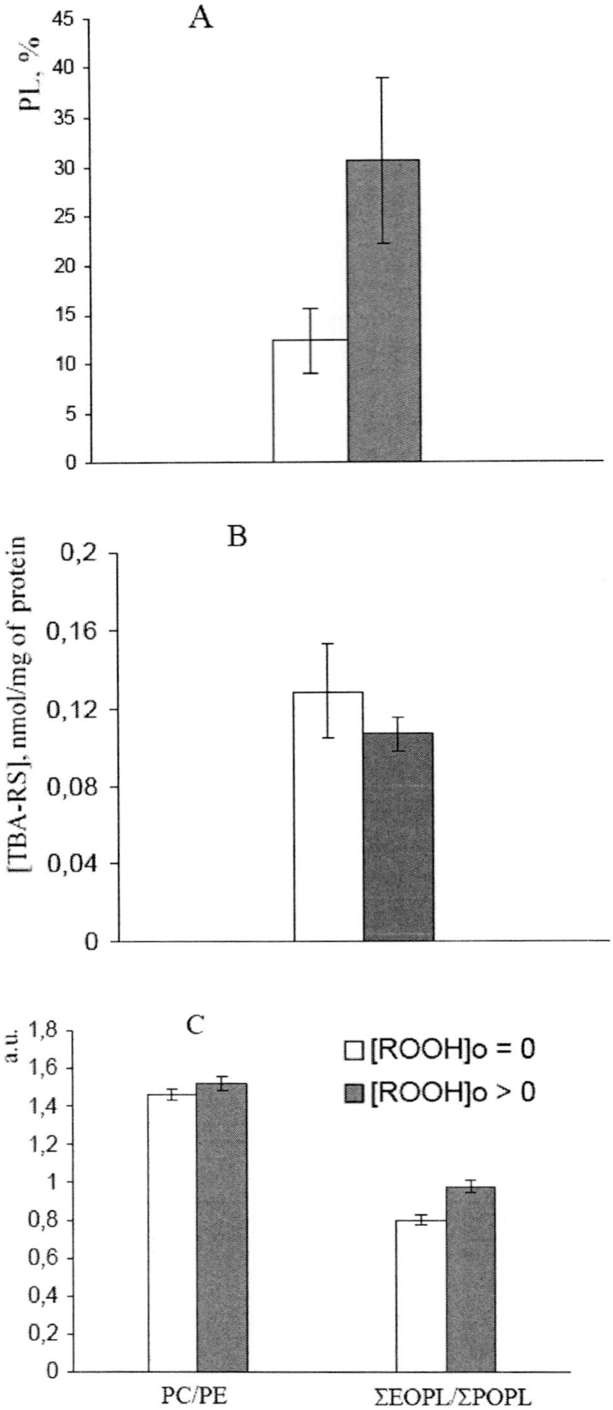

Figure 1. Parameters of the lipid peroxidation in liver of the control mice: □ - the liver lipids exhibit an antiperoxide activity; ■ - the liver lipids contain peroxides; A – the phospholipids proportion in the total lipid composition; B – the amount of the TBA-reactive substances; C – the ratio of phosphatidyl choline to phosphatidyl ethanolamine and the ratio of the sums of the more easily to the more poorly oxidizable fractions in phospholipids.

The dependence of the scale and/or direction of correlations on the lipid kinetic characteristics was revealed in the study on the interrelations between lysoforms of PL and PC amounts and the ratio of the more easily and the more poorly oxidizable fractions and the PC/PE ratio in PL liver in the control group of mice. Thus, the interrelation between the lysoform relative content and the PC proportion in the murine liver PL whose lipids have the APA properties is practically linear ($R = 0.999$) with the linear regression coefficient 1.53 ± 0.03 (figure 2,A). In mice, whose liver lipids contain peroxides, the similar correlation is reciprocal ($R = -0.90 \pm 0.08$) with the linear regression coefficient 1.1 ± 0.2 (figure 2,A).

Regardless of the presence or absence of peroxides in their liver lipids the reciprocal linear correlations between the ratio of the more easily and the more poorly oxidizable fractions and the PC/PE ratio in PL liver in the control age group of Balb/c mice were revealed. However, because of the significant decrease in the proportion of the more easily oxidizable fractions in the murine liver PL when lipids exhibit the APA properties (figure 1) the experimental data are positioned on different correlation lines (figure 2,B) Furthermore, the certain differences are determined for the coefficients of the linear regression of the correlation between the lipid ability to oxidation and the structural state of the membrane system of liver. For the control age groups of mice the coefficient of the linear regression of the reciprocal correlation between the parameters of ΣEOPL/ΣPOPL and PC/PE is equal to 0.516 ± 0.065 ($R = -0.98 \pm 0.02$) for lipids with APA properties and to 0.324 ± 0.075 ($R = -87 \pm 0.11$) for lipids containing peroxides.

In the experimental group of mice the liver lipids as rule contain peroxides. APA lipids was found only in liver of one mouse. The analysis of consequences of the single administration of the surfactant and acetone at the low doses to Balb/c mice was performed relative to the same parameters of LPO process for the control group of mice whose liver lipids also contained peroxides.

It should be noted that the extent of changes in the various average parameters of LPO in liver of mice is different in 1 month after administration of the 0.3 % Tween 80 in the 10 % aqueous acetone solution. The average values of the content of the TBA-reactive substances in liver homogenate, the amounts of the peroxides in lipids and the PL proportion in the total lipid composition do not differ significantly in the experimental and control groups due to the high individual variability of these parameters (figure 3,A; 3,B). The average magnitude of the ratio of the sums of the more easily to the ore poorly oxidizable fraction is the significantly lower and the PC/PE ratio tends to increase for the liver PL in the experimental group as compared with that in the control group (figure 3,C).

The administration of the solution of the 0.3 % Tween 80 in the 10 % aqueous acetone to Balb/c mice causes the disturbance of biochemical and biophysical parameters in tissues that is revealed in a change in the scale and direction of interrelations between the LPO regulatory system parameters. The absence of the correlation ($R = -0.10$) between the content of lysoforms and the relative content of PC in the murine liver PL in 1 month after the action of the chemical agents was shown (figure 2,A). The reciprocal correlation between the lipid oxidizability and the PC/PE ratio retains ($R = -0.998 \pm 0.002$) in the experimental group of mice, however, coefficient of the linear regression (0.864 ± 0.025) is 2.7 times higher than that for the control group (figure 2,B).

The analysis of the interrelation between the primary and secondary products of oxidation in the murine liver in the experimental and control groups showed great changes in the function of the murine liver after injection of the surfactant and acetone at low doses (figure 4). While the direct correlation (R = 0.96 ± 0.03) between the content of the TBA-reactive substances in the homogenate and the amount of peroxides in liver lipids was revealed for the control group, the correlation was reciprocal in the murine liver for the experimental group (R = – 0.77 ± 0.21).

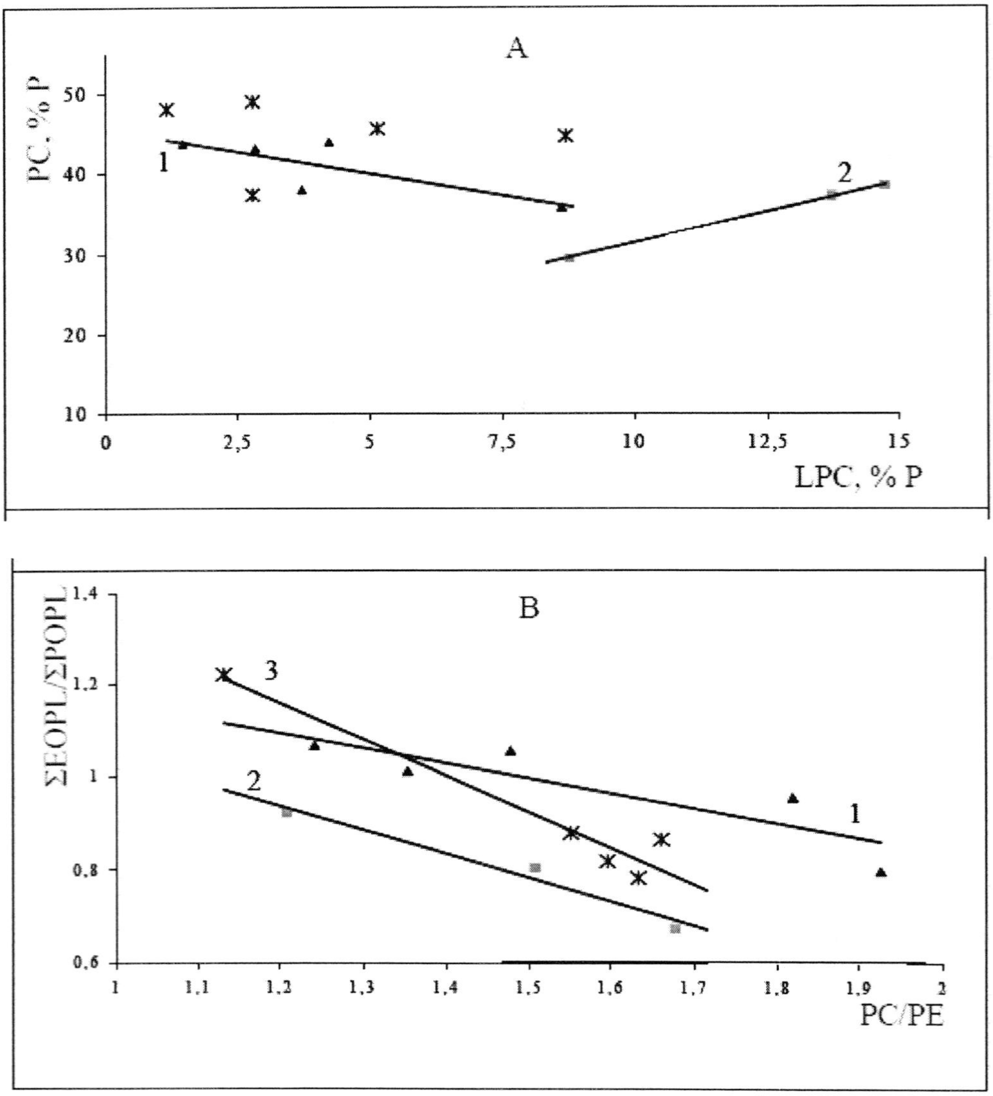

Figure 2. Interrelation between the proportion of phosphatidyl choline and the relative content of lysoforms of phospholipids (A) and between the ratio of the sums of the more easily to more poorly oxidizable fractions and the ratio of phosphatidyl choline to phosphatidyl ethanolamine (B) in the liver phospholipids.: 1 – the liver lipids contain peroxides (the control group of mice); 2 – the liver lipids exhibit an antiperoxide activity (the control group of mice); 3 (*) – the liver lipids contain peroxides (the experimental group of mice, i.e. in 1 month after the Tween 80 and acetone administration.

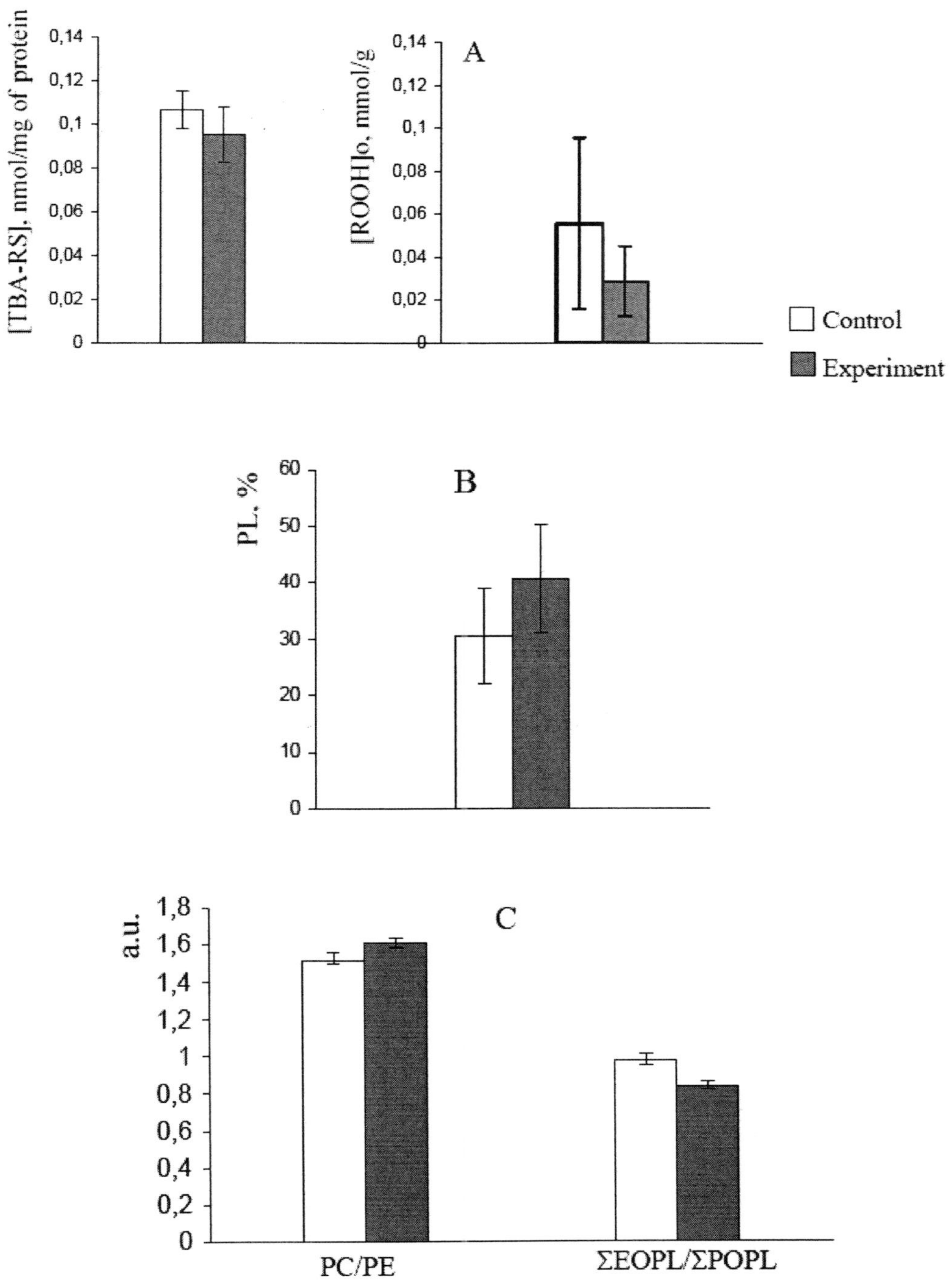

Figure 3. Various parameters of the lipid peroxidation in liver of mice in the control and experimental groups: A – the content of the TBA-reactive substances and the amount of peroxides in lipids; B – the phospholipid proportion in the total lipid composition; C – the ratio of phosphatidyl choline to phosphatidyl ethanolamine and the ratio of the sums of the more easily to the more poorly oxidizable fractions in phospholipids.

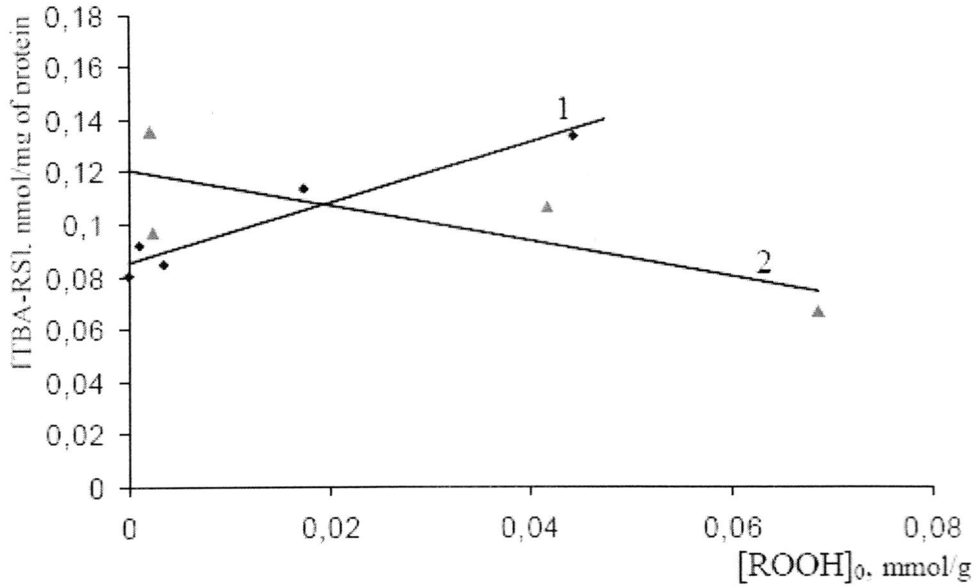

Figure 4. The influence of the intraperitoneal administration of Tween 80 and acetone on the interrelation between the content of the TBA – reactive substances in the homogenate and the amount of peroxides in the liver lipids of the control mice (1) and in 1 month after the action (2).

The data obtained lead us to conclusion about the substantial role of the kinetic characteristic of lipids for the function activity of the murine liver and the persistence of the disturbances in the intensity of the oxidative process and the disbalance in processes of the biosynthesis and degradation of PL a long time after the action of the low-toxicity chemical agents at the low doses on the animal organism.

This work was support by the Foundation of the Program of Presidium of Russian Academy of Sciences “Basic sciences – for medicine”.

REFERENCES

[1] A.G. Kudyasheva, L.N. Shishkina, N.G. Zagorskaya, A.I. Taskaev, “Biochemical mechanisms of the radiation damage of the natural rodent population”, St.-Petersburg, *Nauka,* 1997, 156 pp. (In Russian)

[2] E.V. Shtamm, L.N. Shishkina, N.B. Kozlova et al. *“Vodosnab. and sanitar. tekhnika”,* N 10, pp. 18 – 21, 1997 (in Russain).

[3] T.I. Bonashevskaya, N.N. Belyaeva, “Morphofunctional evaluation of the genotoxic action of the chemical environment factors”, pp. 20 – 25, in the book *“Structural – functional and biochemical mechanisms of effect of the environmental factors on humans and experimental animals”,* Moscow, 1986 (in Russian).

[4] V.V. Urnisheva, L.N. Shishkina, *Biophysics* Vol. 48., N 4., pp. 744 – 753, 2003.

[5] V.V. Urnisheva, L.N. Shishkina, *Bull. RAS, Ser. Biology,* N 2, pp. 163 – 168, 2004.

[6] Frankfurt, Voprosy onkol., Vol. 21, N 10, pp. 63 – 67, 1975 (in Russian).

[7] E.B. Burlakova, A.V. Alesenko, E.M. Molochkina, N.P. Pal'mina, N.G. Khrapova, "Bioantioxidants in the Radiation Damage and the Tumor Growth", Moscow, *Nauka,* 1975, pp. 214 (in Russian).

[8] M. Kates. The technique of lipidology. Moscow: *Mir,* -1975. 322. pp. (Russian version).

[9] 9. "Biological membranes. A practice approach." / Eds J. B.C. Findlay, W.H. Evans – Moscow: *Mir,*-1990. 424 pp. (Russian version).

[10] T. Asakawa, S. Matsushita. *Lipids.* - 1980. - V. 15. - No. 3. pp. 137-140.

[11] R. Itzhaki, D.M. Gill. A*nal.Biochem.*- 1964.- V. 9. pp. 401 - 409.

[12] G.F. Lakin. Biometry. 3rd publication. Moscow: Higher school, 1990. 293 pp. (in Russian).

[13] E.F. Brin, S.O. Travin, *Chem. Phys. Reports,* Vol. 10, N 6, pp. 830 – 837, 1990.

[14] O.K. Khmel'nitskii, V.L. Bykov, *Arkhiv patolog.,* Vol. XIII, N 6, pp. 3 – 9, 1986 (in Russian).

In: New Aspects of Biochemical Physics
Editors: Sergei D. Varfolomeev et al., pp. 123-135 ISBN 1-60021-419-3

Chapter 17

FREE-RADICAL MECHANISM OF CHITOSAN RADIATION DEGRADATION AND PROBLEMS OF CELL PROTECTION

O.A. Pilipchatina and V.A. Sharpatyi
N.M. Emanuel's Institute of Biochemical Physics Russian Academy of Sciences;
4 Kosygin str., Moscow 111999, Russia; E-mail: vashar@sky.chph.ras.ru

ABSTRACT

An analysis of published data on the radiolytic properties of chitosan and some modelling its fragments substances was performed. The main process of radiation chitosan degradation is connected with the formation and conversion of free radicals. The mechanisms of the primary radicals $\dot{C}2$, $\dot{C}1$ and $\dot{C}3$ (of the NH_2 and H removal radicals) conversion up to the formation of some end products of chitosan radiolysis were offered in the form of the generalized schemes. Problem of the DNA and membrane protection in irradiated cell is discussed.

Keywords: Chitosan; mechanism of radiolysis; free radicals, scavengers; radioprotection of DNA and membrane in cell.

INTRODUCTION

Chitosan is high-polymer of glucosamine. Chitosan of molecule mass from 10 up to 70 kDa is an effective radioprotector: the intravenous injection of it into mouses befor the irradiation increases the animals survivability up to ~75 % [1]. Chitosan is produced enzymically or by the concentrated alkali treatment from the chitin – the natural analog of cellulo-se, consists of the N-acetyl-α-D-glucosamine residues [2]. The application chitosan as a radio-protector in comparison with others is more preferable because of biocompa-tibility and biodegradation up to glucosamine or N- acetylglucosamine (in the not completely desacetylic drugs) [3] – metabolites of an organism animal and man.

On Free-Radical Mechanism of Chitosan Radiolysis

Despite of more than 40-year's history of radiobiologic investigations on chitosan [1] and available data on it radiolytic properties, generalized in [4,5], complete pattern about the mechanism of a protective action of this polymer compound, as well as about the mechanism of its radiation destruction, still is not present. The literary data of last years, including investigations on radiolytic properties of polysaccharides and low-molecular carbohydrates simulative the chitosan and chitin fragments, allow us to present the basic directions of their radiation destruction on free-radical mechanisms – sequence of the elementary acts of the primary macroradicals transformations down to formation of some molecular products of radiolysis.

In the present review the results of the radiolytic properties study of the substances – chitosan precursors and substances modelling its separate fragments as of the polymer compound, and transformations of free radicals in these systems are analysed.

Radiolysis of Glucosamine (GA)

A attribute reference of the GA radio-lytic properties in water solutions distinguishing from a radiolysis of neutral carbohydrates, is the interaction it with a hydrated electron: $k_{e+GA} = 4.2x10^8$ l $mole^{-1}$ c^{-1}. The participation of the electron is demonstrated by the results of ESR-measuring of the atom H yield in irradiated at 77 K phosphate glass-like GA solutions of different concentrations. The decrease of H yield with elevation of GA concentration reflects passing competing processes:

$$e + H_2PO_4^- = H + HPO_4^= \quad (1)$$

$$e + \text{Glucosamine} = \text{Products} \quad (2)$$

Radicals of Glucosamine

Under the OH and H action on a GA as well as in case of neutral carbohydrates the removal of atom H from all of six atoms C is possible The glucosamine residues in the chitosan and chitin mo-lecules are present in the cyclic form. It is characteristic as well as for GA-hydrochloride (GA.HCl), and N-acetylized glucosamine (N-Ac-GA).

At the low-temperature irradiation of the GA.HCl the radicals are formed (with a prevailing yield), ESR spectrum of which is a doublet with a ~ 2,6 mT (R_d), caused by interaction of unpaired electron with the proton on C2. At heating of the sample approximately up to 170 K in a spectrum a new wide (ΔH ~ 8 mT) line with five equidistant (~ 2 mT) components arises, interrelation of which intensities, is close to binomial, – quintet (R_q). The change of intensity of the doublet and quintet components, reflecting concentration of R_d and R_q radicals, follows antibate dependence, and the summary concentration of radicals in intermediate temperature range is maintained by a stationary value in limits of errors of measurings. From here it was made the deduction about the transformation of radicals R_d into

R_q through the isomerization of R_d with the opening of the residue cycle by break of C5–O5 bond: the atom C5 has an unique place in the radical, where unpaired electron can interreact with the four equivalent protons.

The HFS of the irradiated N-Ac-GA spectrum differs from the HFS of GA and GA.HCl spectra, that is caused by the appearance of the substituent in a N-position and by the formation of radicals in it. The feature of transformation of radicals in N-Ac-GA at the annealing is the formation of allyl-type radicals such as $(-\ C(=O) - \dot{C}H -)$. The parameters of this radical spectrum (doublet with $a \sim 2$ mT, g=2,007) are caused by a delocalization of a unpaired electron by the fragment of carbonyl group. The formation of allyl-type radicals, as well as at the irradiating of neutral carbohydrates is connected with the primary hydroxyalkyl radicals transformation owing to the H_2O elimination.

Radiolysis of Acetamid

N-substituent, simulative a fragment, in chitin. In irradiated at 77 K acetamid (H_2N–CO–CH_3) the anion - radical (electron attacks > C=O group; a doublet 1:1, $a \sim 2,9$ mT, g=2,003). Under activity of γ- radiation it decomposes with the acylic radical formation (ΔH = 0,4 mT, g = 2,0009) [6].

So, from the ESR-measurements on chitosan and chitin low-molecular models follows, that the radicals transformation in the N-replaced unhydroglucose cycle can result in to breakages in a polymer chain and to destruction of monomer units. The participation of an electron is reduced to responses with amino-groups and N-substituents.

Radiolytic Properties of Heparin

Heparin, a sulfated glucosaminoglu-can, polymer compound of glucosamine and uronic acid residues, connected with glycoside bonds; polyanion; sulfa- and carboxygroups in heparin are the charge carriers (chitosan is polycation). The drugs of heparin with 20-100 monomer unities are applied in medicine as an anticoagulant of blood. By sulphonic-acid groups of a pentasaccharide fragment with a particular sequence of residues (N-acetyl-α-D-glucosamine-6-sulfate; β-D-glucuronic acid; α-D-glucosamine-N-sulfate-3,6-disulfate; α-L-iduronic acid-2-sulfate; α-D-glucosamine-N-sulfate-6-sulfate). Heparin contacts to plusly charged Lis and Arg residues in antithrombin III, and an Arg residue contacts with the second blood plasma protein Xa (coagulative factor Xa). The ternary complex is frameed, in which pentasaccharide inhi-bits process of formation of a fibrin, i.e. interferes with coagulation of blood [7].

Under the irradiation heparin in water solutions and in a dry view loses anticoagulative properties [8, 9]. In irradiated solutions of heparin hexauro-nic and glucosamine residues are destroyed, the content of nitrogen, sulpho-nic-acid groups is reduced (desulfatation), the molecular mass decreases (depolymerization), the carbo-grours amount in polymer grows and acidity of solution grows as well [8]. The destruction of heparin in solutions, the formation of carboxyl-groups in it descends under influence of OH-radicals ($k_{OH+heparin} = 4.7 \times 10^8$ l $mole^{-1}c^{-1}$); the e_{aq} reactivity to heparin is less on an order of magnitude [10]. e_{aq}. participates in process of heparin depolymeri-zation, but it, in opinion of the authors [10], is not accompanied by losses of sulphonic-acid groups in polymer.

In heparin water solutions (8-30 %), irradiated at 77 K the radicals of heparin, and of water (OH and $e_{st.}$) are generated, G (e_{st}) ~ 0.1[11]. This value is much more less than one apparent in the 30 % -s' solution of D-glu-cose irradiated in the same conditions: G (e_{st})=0.9. Hence it follows, that at least 90 % of electrons at 77 K react with N-substitutes in glucosamine (in-clouding heteroatoms and olefinic linkages), just as it takes places at a low-temperature irradiation of amino-sugars, peptides and proteins [6]. With the count of the data [8] we concluded also, that the processes of desulfatation and the decreasing of the nitrogen amount in heparin are initiated by e.

ESR-spectra of heparin solutions at 77 K are a result of superimposition of basic doublet (1:1) and triplet (1:2:1), centered in region g = 2.003 [10]. As well as in case of carbohydrates [6,12], doublet is from the radical C1 (splitting on proton). The triplet (splitting on two protons) can belong to the radicals of a removal of H from C2, C3 and C4. The similar spectra (doublet + triplet) in irradiated heparin were observed also in [9,13]. The HFS spectra constants are conditioned by the interaction of an unpaired electron with protons: a = 2.05 mT and a = 2.95 mT [13]. At the heating of irradiated samples up to T = 323 K these radicals are disappeared and only the ions-radicals SO_3^- are registered (the broad single line [9]). The appearance of heparin ions-radicals at 77 K shows the participation of electrons in desusfatation process.

In the case of irradiation of the solid state complexes of heparin –deter-gents (of various composition: cetyl pyridinium chloride, CPC, cetyl dime-thylbenzylammonium bromide and cetyl trimethylammonium bromide) the effective protection of heparin from destruction by the aromatic moieties of complexes is found [14]. So, in the heparin-CPC sample, irradiated at ambi-ent temperature, the radical yield reaches G = 0.2, the value is more than 10 times smaller in comparison with registered one in the components of the complex irradiated individually. ESR-spectrum of the irradiated at 300 K complex heparin-CPC belongs to H-adduct radical of the pyrimidine moiety.

In these conditions of experiment the formation of carboacids and aldehydes, inorganic sulfates, decrease of molecular mass of heparin are not filed. The numbered experimental facts testify to effective protection of heparin in a complex by its aromatic fragment, which can be explained in terms of the free-radical mechanism of energy transfer from heparin to an aromatic moiety of the complex.

Being returned to a main parameter of a functional biological potency of heparin, its anticoagulative properties (parameter showing of integrity and molecular, and supermolecular structures of a polymer compound), we shall note, that filed at an irradiation of heparin the losses of anticoagulative properties can be connected, in our opinion, with 3 factors activity. The main of them is losses of heparin sulphate groups (under influence of an electron) and from here the impossibility of formation of a ternary complex "pentasaccharide - antithrombin III -Xa ". The 2-nd factor is the change of a conformation of heparin chain because of formation of epimer in any of the pentasaccharide residues (the conformity, complement of a locating of heparin anions and plusly charged amino-acid residues is broken). The epimers appearance process on response $R + e = R^-$ in the macromolecules with carbohydrate moieties (for example, in DNA) surveyed in [15]. The3-rd factor is the transformation of any primary radical in pentasaccharide with the breakage appearance in a macromolecule, that also interferes with formation of the specified complex.

CHITOSAN RADIATION CHEMISTRY

The main processes of chitosan destruction are: deamination (G ~ 3), formation of breaks in polymer chain (G = 4.2 ± 0.3), abjection of molecular hydrogen (G = 1,3) and dioxide of carbon (G = 1,5) [4]. In irradiated at 77 K chitosan and chitin ESR method registered the radicals. The sum yield of radicals at 77 K in chitosan is 3.0, in chitin G(ΣR) = 2.3.

In case of the chitosan irradiation at 77 K, in which ~ 30 % of amino-groups are replaced by the quaternary ammonium groups (chloride ammoni-um), the radicals yield is G = 1.9, the of anions-radicals Cl_2^- yield is G = 0.04. At the chitosan and trioxybenzoic acid (3,1 % of masses.) conjugate irradiation at the same conditions the yield of radicals decreases up to 0.76, and the yield of ions – up to G = 0 [16]. These data specify that, as well as in surveyed earlier system "heparin – aromatic compound (AR)", at the com-plexes chitosan-AR irradiation the transfer of energy, immersed by a poly-mer, the second moiety of this complex also is effective. Hence it appears, that if heparin and chitosan are used in radiation medicine, that for the radi-oprotection of both polymers it is necessary to bring in their composition "small quantities" of the "AR".

Mechanism of Chitosan Radiolysis

The main difference of aminosugars radiolysis from radiolysis of neutral ones is the fact that they (aminogroups) react with electron. The result of the interaction of **e** with aminogroups of chitosan ($k_{e+chitosan} = 4.6x10^8$ l.mole^{-1}c^{-1} [5]) is the deaminization of the poly-mer [39]: alkyl radicals C2 and ammonia are formed. The similar processes are revealed in proteins, irradiated at low temperature, – deamination of Lis aminoacid residues [6]. In irradiated at T = 300 K chitosan carboxyl- (G ~ 0.6) and carbonyl groups are accumulated, and the yield of reducing groups is 25 ± 2 (it was determined by copper number) [4, 17,18]. The mechanisms of these radiolytic products formation are analogous to the ones examined recently for the neutral polysaccharides radiolysis [6].

The Free-Radicals Structure and Transformations of Them

By ESR spect-ra in irradiated at low temperature chitosan and chitin the the primary radi-cals of H removal from C1 (doublet 1:1, *a* ~ 3,0 mT) and from C4 (triplet 1:2:1, *a* ~ 2,6 mT) were identified. Under the heating of the irradiated chitin (acetylated chitosan) up to T ~ 200 K the concentrations of $\dot{C}1$ and (essen-tially) $\dot{C}4$ were decreased and the spectrum of radical GluNH$\dot{C}$(OH)CH_3 (1:3:3:1, *a* ~1.4 mT) was registered, where Glu was the fragment of polymer chain [17, 18]. Analysis of presented in [17, 18] ESR spectra of irradiated chitosan (figure 1) and chitin allows us to determine supplementary the line with the total width ΔH ~ 9,5 mT registered at 77 K against a background of others (and for some reason not taken into account by the authors). We inter-pret it as doublet of triplets (a_d~ 2 mT, a_t~ 3,8 mT, ΔH_t~ 1 mT, g ~ 2,003) attributing it to the radical of the aminogroup from C2 removal. Concentra-tion of these radicals at 77K is 20-25 % from total amount of radicals [19].

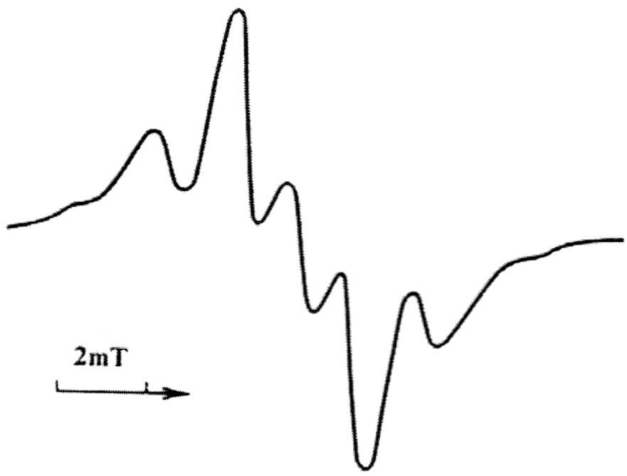

Figure 1. ESR spectrum of the γ-irradiated at T = 77 K chitosan [17].

The authors of work [5], studying the chitosan radiolysis in aqueous solutions (pH ~ 3, packed by N_2O, the reacting agents are the radicals OH), specify an opportunity of the formation of radicals of H removal from all atoms C in the unhydrocarbohydrate cycle. They explain the breaks in a polymer chain formation by the transformations of the primary hydroxyalcy-lic radicals C1, C4 and C5 [20,21]. For the radicals C1 and C4 two variants of such transformations are supposed: a hydrolysis of glycoside bonds in the radicals C1 and C4 at the β-position and the radical C1 fragmentation at the β-bond position with the lactone formation and transmission of unpaired electron on the next monomer unit. The transformation of radical C5 accom-panied by disclosing of unhydrocarbohydrate cycle with appearance in it carbonyl groups on C5 and C1 and the jumping of unpaired electron on C4 of the next monomer unit. At the presence O_2 α-hydroxy-peroxy-radicals C2, C3 and C6 are arised, transformation of which accompanies by the elimina-ting HO_2 with the carbogroups formation on these atoms C. It is underlined, that aminogroups of chitosan during the radicals transformations behave similarly to OH-groups [5].

Radicals C1 may transform by two directions - isomerization of $\dot{C}1$ with break of C5-O5 and carry-out of unpaired electron on C5. Such process is registered by ESR method at low-temperature irradiation of glucosamine [22]. Further stages of transformations of radicals C5 in the presence of O_2 for example in dry starches lead to formation of malic or tartaric acids [21].

The other one direction of the $\dot{C}1$ transformations is isomerization with break of a glycoside bond $\dot{C}1$-O1-C'4 and formation of keto-group at C1 (lactone) with unpaired electron carry-out to C'4 of neighbouring monomer unit (as well as is supposed in [5]). With the transformations of a radical $\dot{C}'4$ at presence O_2, the accumulation of compounds of 2-amido-2-desoxytetradi-aldose (from glucosamine residue) and 2-acetamidodesoxytetradialdose (from N-acetylsubstituted residue of glucosamine) type is connected. The last product was obtained under irradiation at presence O_2 saturated N_2O solutions of N-acetyl-2-glucosamine (G = 0.5, 0.01 M [12]).

At the presence of water in the radicals $\dot{C}1$-O1-C'4 the bond O1- C'4 can be hydrolyzed [5,23], and at the atoms C1 and C'4 in the next monomer unit the hydroxyl groups are appeared. During the transition of the H from hydroxyl C1OH to O5 an acyl-radical is formed (on C1), eliminating CO; CO in chitosan was registered [12,17].

For radical C2 three variants of transformations are possible [19]:

a) by the type of water β-elimination (removal of H from C3 and OR from C4). The break appears in polymer chain and the desoxy (C4) - and keto (C3) -groups are formed in monomer unit at the same places as under citric and siccine acids formation in starches [21];
b) isomerization with break of bond C3-C4; by the formation of double bond C2=C3 and radical with non-coupled electron at C4 under radiolysis of glucosamine in solutions the glycerin aldehyde was registered (in 0,05M solution G = 1,3 [12]);
c) regrouping of bonds H–C at C1 and C2 with carry-out of non-coupled electron on C1 and further transformation of radical C1 analogously to radical C2 transformation in the case of cellulose [19,20] Subsequent ones the compounds are presented identified under radiolysis of glucosamine or N-acetylglucosamine [12].

The proof of an opportunity of embodying of variant "a" is served by detection among products of the starch radiolysis of the citric and succinic acids keeping at atoms C3 and C4 the specified functional groups.Sequence of the elemenatary acts of this process in details is surveyed in [20, 21].

The transformation of the alkyl radical C2 on a route "b" would prove to be confirmed by findings of investigation of a radiolysis of glucosamine in solutions (0.05 M). In this case glyceraldehyde is registered, G = 1.3 [12].

The first stage of the radical C2 transformations on a route "c" is redistribution of the H-C bonds at atoms C1 and C2 in a cycle, that is similar to that descends in case of cellulose [19]. The radicals C1 can transmute on two directions - isomerization with a breakage of C5–O5 bond and run-over of a unpaired electron on C5. Such process is filed by a method ESR at a low-temperature irradiating of glucosamine [6,22]. The subsequent stages of C5 radical transformation, for example, in the irradiated in presence of O_2 starches leads to the malic and tartaric acids [21].

If under chitosan irradiation in the presence of O_2 the desamination act is accompanied by formation of peroxide radical C2OO· and then by formation of peroxide (by C2) then under its decomposition with break of bonds O–O and C–C2 (as well as in irradiated starches [24]) by C2 the aldehyde group is appeared. As a result of regrouping of bonds at C1 the arabinose may be formed

If the sample of chitosan contains besides of glucosamine residues the N-acetyl-β-D-glucosamine(not fully deacitylated chitin moieties), we should take into account an opportunity of electron capture by carbonyl group of acetyl with formation of anion - radical. The similar process is observed in case of the low temperature irradiation of acetamide, of proteins and pepti-des: electron attacks carbonyl group of peptide bond. Identifiable at 77 K in the irradiated proteins anion-radicals (poor-resolved doublet with $a \sim 1.5$ mT) at the sample heating up to 190 – 200 K is protonized (at interaction with molecule of water finding mobility owing to structural change). The neutral radical - H-adduct with the approximately same ESR-parameters is formed, as well as in the initial anion- radical (poor-resolved doublet R_p [6]).

Parameters of the ESR-spectrum and conditions of chitin irradiation and the GluNHĊ(OH)CH_3-radcial registration [4,17] allowed concluding that in given system the H-adduct of acetyl from anion-radical is formed. Hydrolysis of N-C in this radical is accompanied by the formation (with equal yields) of free amino-group and molecule of acetic acid [4]. Such interpretation of effect of deacetylation in chitin [4,17] is seemed for us more acceptable.

The schemes reflect the participation of radicals appearing under the action of electrons. With consideration of the values of yield of chitosan de-struction process (formation of breaks) and deamination yield we may assu-me that more than 2/3 of effect (71 %) is fallen at the last one, and consequ-ently at the participation of electrons in polymer destruction. The rest 29 % reflect the contribution into destruction of primary radicals of removal of H.

Investigations of radiation chemistry of glucosamine and N-acetylglucosamine were carried out at solutions concentrations 0,01-0,1M, i.e. at conditions of total involving of forming from water radicals into reactions with dissolved substances [12]. Electrons attack the fragment N-C2, and radicals OH attack H–C bonds of all 6 carbon atoms, radicals of H removal are formed [18]. The last fact is confirmed by the structure of identified products of radiolysis in the absence of O_2, i.e. by compounds containing keto- (or aldehyde) group at each of carbon atoms. Mechanisms of formation of these compounds are considered in [12].

The schemes (1, 2) are given on the basis of the dry polysaccharides ra-diolysis examinations. They are applicable and for a liquid-phase radiolysis, as and in this case aminogroup leaves under influence of an electron.

Analysis of a radiation chemistry of glucosamine and N-acetylglucos-amine in aqueous solutions were carried out at concentrations 0.01–0.1M,i.e. in the conditions of complete recruitment phenomenon of radicals, gene-rated from water, in response with solutes (the effect of an indirect action of radiation is peak)[12]. The electrons attack a fragment N–C2, and the OH at-tack the bonds H–C of all 6 atoms C, the radicals of H removal are formed [6]. The last was proved by the composition of identified products of radio-lysis (in absence of O_2) – substances keeping keto- (or aldehydic) groups at each atom C. The mechanisms of formation of these bonds surveyed in [12].

Under irradiation of dry chitosan the value of yield is $G(H_2) = 1.3$, and. it may be used as basic one under estimation of contribution of radicals of H removal into total process of polymer destruction, if we remember that H_2 in polysaccharides is formed by the reaction $^{\cdot}H + HR = H_2 + {^{\cdot}R}$ [17]. Then the part of H removal radicals in chitosan destruction is 1.3 / 4.2 ~ 0,3. so, we get the same value equal to 30 %, as well as that one determined above by the difference between the yield of sum destruction process and contribution into this process of electrons conditioning desamination of polymer. Direct measurement of radicals concentration in chitosan at low temperature gave the value G = 3.0 [4]. This value exceeds the estimation one, but it is lower if we assume, that one radical causes one break in a polymer chain.

Formation of CO_2 in irradiated chitosan (or chitin), keto- and desoxy-ketoderivatives of glucoamine in solutions of chitosan, glucosamine and N-acetylglucosamine is connected with transformations of radicals of H remo-val. The scheme of the mechanism of CO_2 formation in the absence of O_2 (that means that by C1, since only it is connected with oxygen atoms [23]).
i

Formation of arabinose is connected with transformations of radical C1 in the absence of O_2. Mechanism of this process for solutions of glucosami-ne and N-acetyl-substituted glucosamine is considered in [12]: after hydrol-ysis in ·C1 of bond β -O1–C'4 and transition of H on O5 the acyl radical C1 eliminated CO, and in the case of N-acetylsubstituted glucosamine the radi-cal –Ċ2HNHAc is formed. Under its interaction with water acetamide is removed and end-hydroxyradical is formed, oxidation of which (removal of H from hydroxyl) leads to formation of arabinose G (arabinose) = 2.1.

Formation of keto- and desoxyketosugars of N-acetylglucosamine [12] with consideration of data on cellulose and starches [21,24] represents multi-stage process of transformations of the H removal primary radicals. For example, ·C3 transformations include the following stages: β-elimination of water (with participation of cyclic atom O [28]), bonds regrouping in allyl type radical, isomerization with bonds C1–C2 and O–H breaks in hydroxyl at C1 and removal of CHO. The last one is dimerized with forma-tion of glyoxal registered under cellulose and starches irradiation.

Process of formation of desoxyketogroup –C4H_2–C3(=O)– in cycle of N-substituted glucosamine is presented in scheme 4b [24]; in polymer the hydrolysis of β-bond in glycoside radical C1 and break of C1–C2 bond with carrying-out of H from hydroxyl C1 on atom C2 occur. Further hydrolysis of bond C5–O5 in radical leads to 1-acetamido-1,3-didesoxypent-2-ulose regis-tered in irradiated 0,01M solution of N-acetylglucosamine (G = 0.1 [12]).

In conclusion we should note two moments:

a) 30 % of effect of chitosan destruction in the absence of O_2 fall at the part of primary radicals of removal of H, and the value of yield of only one product CO_2 equal to 1.5 even slightly exceeds these 30 %. From the other hand, the yield of radicals in chitosan is lower than destruction yield. We may suggest that some of radicals while transforming transmits unpaired electron to neighbor monomer units and initiates their decomposition. For example, in the case of cellulose it was found that as a result of radical C1 transformation 2 neighbor monomer units are decomposed due to transfer of non-coupled electron from C1 to C'4. In the upshot we obtain that without change of sum amount of radicals in system the amount of products of their transformations is increased. Mentioned value of yield of formation of redu-cing (carbonyl) groups in irradiated chitosan (~ 8 carbonyl groups per 1 radi-cal) serve as confirmation of the last fact. Destruction of neighbor monomer cycles doesn't influence on number of breaks in polymer;
b) chitosan destruction is caused by participation of both the electron and radicals OH and H. It means that chitosan under irradiation of aqueous solutions may effectively capture the water's radicals[25]. According to [26], chitosan is effective scavenger of alkyl, hydroxyl and superoxide radicals. Considering the structure of linear polymer, the presence of two hydroxyl and one amino groups in monomer unit which may form in water molecule hydrogen bonds as it is proved in the case of starch (structuring of water by it) we want to note that chitosan obviously possesses the same properties. Its high activity as radioprotector is probably caused by ability of polymer to structure the water (gelatinization) and consequently to scavenge the water radicals " at distance "[27]. At intravenous introduction of chitosan of molecular mass from 10 up to 70 kDa to mouses (from calculation 20 mg/kg for 15-30 min) before the irradiation in dose 8 Gy (minimum absolutely lethal) increases the animals survivability up to 73 % [1]. This fact may serve as illustration of

radioprotection by mechanism of competition for radicals of water, and revealing of effect of indirect irradiations action.

On Mechanisms of Cell Protection by Chitosan

Being returned to a problem of radioprotection of a cell, organism, by chitosan and taking into acount of embodying of the indirect action mecha-nism of ionizing radiations, we shall make attempt to present chitosan as DNA and membranes (two basic cell targets) protector [28].

Chitosan in a Role of DNA Protector

By experience in vitro [29,30] shown, that DNA and chitosan can form complexes. Citosan-DNA nanopar-ticles of the size 100-250 nm and of the composition 35.6 weight % DNA and 64.4 % chitosan are formed owing to electrostatic interaction between amino groups of ligand and phosphate groups at ratio N/P = 3 - 8 (at chito-san concentration of 100 microg/ml) [31]. The formation of similar particles with a ratio N/P = 4 and the size of 400-600 microns (200 microg/ml) is re-vealed and on the other system - at preparation of an anticaries DNA vaccine (plasmid pGJA-P/ VAX) in chitosan [32]. The transfection efficacy depends on a type of used cells. The opportunity of chitosan penetration through cy-toplasmic membrane of various cells is scored also in [33]. If chitosan insi-nuates into a cell and forms with DNA a complex, as in vitro [29,30], being in a composition of complex it could represent itself as the free radicals sca-venger,similarly to that descends in a case of sulfurcontaining amines– radi-oprotectors [34]. (The latter ones are attached with the amino groups to ne-gative charged atom of oxygen in DNA phosphates[35].) Opportunity of formation of complexes"sulfurcontaining amine – DNA" and scavenging of free radicals is confirmed by apparent decrease of a mutation frequency in irradiated animals: mouse, rat, rabbit (injected with the radioprotectors) and Drosophila (radioprotector diet) [36].

It is possible to figure, that the basic mechanism of chitosan penetration in a cell as well as for other macromolecules (proteins, lipids, glycoproteins) is endocytosis. The sizes of chitosan macromolecules in a drug of maximal efficiency (at MM = 10-70 kDa [1, 37]) are compounded with the typical ve-sicle dimensions (from 50 up to 1000 nm) for polymer molecule at endocy-tosis . The times of chitosan drugs introduction into animal before the irradi-ation for the achievement of maximal effect of radioprotection [1] correlate to the time of the specified macromolecules penetration into a cell under en-do-cytosis and of a ligand–receptor associates formation (up to10 min [38]).

On Chitosan Radioprotection of a Membrane

If, as well as in a case of DNA, we talk about avoidance of free radicals action on the main frames of a membrane, it is necessary to take into account two factors. The first one: to what membrane fragments the chitosan molecule attaches; the second one: chitosan protects membrane from what types of damages.

Practically all the cytoplasmic membranes of animal cell contain phos-pholipides with negatively charged heads - phosphatic groups, from here it is necessary to figure, that the chitosan protonated aminogroups could be asso-ciated with them. The degree of interlinking them with membrane phospha-tic groups depends on pH of solution, and on molecular mass of chitosan. The interaction of chitosan with lipids of a membrane is testified by the fact, that in this place exactly the order of lipid bilayer organization is variated [39]. At a low-temperature irradiation of aqueous solutions of lyposomes dispersions the radicals of lipids are formed; the yield of them is essentially reduced at the presence of a radioprotector (sterically hindered phenol, ac-ceptor of free radicals), introduced into lipid bilayer of lyposomes before the irradiation [40]. These data have allowed to assume, that chitosan, being attached to phosphatic heads of lipids of membranes, can represent itself as scavenger of free radicals – water radiolysis products.

Proteins are the second main component of membranes. In dependence on a type of cell the content of them in membranes changes from 25 % up to 75 %. The basic function of membrane integral proteins (interesting for us) is to carry out the transport of matters into cell, which are necessary for its functioning.The system of receptors of lipoprotein and glycoproteins – sub-units in a composition of integral protein checks this process of transport. Under the irradiation the structure and functioning of integral proteins are broken, and through a defective membrane any antigens and toxines can pe-netrate into the cell. By experiments [41] on macrophages of bat cells (RAW 264.7) and others ones [42] was shown, that the chitosan aminoglucose resi-dues were associated with receptors of immunoglobulins specific to manno-se, on a system of the complement (key - lock). From here it is possible to conclude, that chitosan interacting with receptors of the immunoglobulin (the protein fragment of a glycoprotein complex) interferes attack by free radicals and maintains the receptor cystem for the subsequent functioning - recognition of antigens. In this case we deal with the conservation of immunity of an organism supressed, as is known, by the radiation.

Thus, the presence of chitosan on a membrane can prevent (or reduce) the changes of structural organization of a bilayer of lipids and the destruction of receptors under action of ionizing radiations.

References

[1] Il'in L.A.,. Andrianova I.E, Glushkov V.A. // *Radiation biologiya. Radioekolo-giya,* 2004. V. 44. № 2. C. 176 (in Russian).

[2] Stepanenko B.N. Chemistry and biochemistry of carbochydrates (polycuccharides). Moscow : *Vysshaya shkola.* 1978. 256 p.

[3] Tsygan V.N., Zhogalev K.D., Nikitin V.Yu. // *Rynok BAD,*2002.№2.P.1.

[4] Ershov B.G., *Uspekhi khimii,* 1998. V. 67. № 4. P. 353. (in Russian).

[5] Ulanski P., von Sonntag C. // *J.Chem. Soc., Perkin Trans.,* 2000. V. 2. P. 2022.

[6] Sharpatyi V.A. Radiation chemistry of biopolymers. Moscow: *Energoizdat.*1981. 168 p. (in Russian).

[7] Van Boeckel C.A.A., Petitou M. // *Angew. Chem. Int. Ed. Engl.,* 1993. V. 32.№ 12. P. 1671.

[8] Balazs E.A., Laurent T.C., Howe A.F., Varga L. // *Radiat. Res.,* 1959. V. № 2. P. 149.

[9] Kudryavtsev Val.N., Grushevskaya L.N., Yurik T.K., Kabanov V.Ya.. // *Khimiya vysokikh energii,* 1993. V. 27. № 1. P. 41 (in Russian).

[10] Baugh P.J., Morgan R.E., Kershaw K. Phillips G.O. // *Padiat. Res.,* 1971. V. 46. № 2. P. 217.

[11] Jooyandeh F., Moore J.S., Morgan R.E/, Phillips G.O. // *Radiat. Res.,* 1971. V. 45. № 3. P. 455.

[12] Kochetkov N.K, Kudryashov L.I., Chlenov M.A., Radiation chemistry of carbohydrates, Moscow: *Nauka,* 1978 288 p (in Russian).

[13] Baquey Ch, Darnez Ch. // *C.R.Acad. Sc.,* 1973. *Ser. C. T.* 277. P. 139.

[14] Moore J.S., Phillips G.O., Rhys D.// *Radiat. Res.,* 1972.V. 50. № 3. P. 479.

[15] SharpatyiV.A. // *Radiobiologiya,* 1992. V. 32. *Vyp.* 2. P.180 (in Russian).

[16] Aleksandrova V.A., Shilova I.A., Kuzina S.I., Mikhailov A.I. // *Proc. Of IV N.S.* Bakh conference on radiation chemistry. Moscow 1-3 June 2005. P.10 (in Russian).

[17] Ershov B.G., Isakova O.V., Rogozhin S.V., Gamzazade A.I., Leonova E.Yu., // *Doklady Akad. Nauk SSSR,* 1987. V. 295. № 5. P.1152 (in Russian).

[18] Ershov B.G., Sukhov N.L., Nud'ga L.A.,. Baklagina Yu.G, Kozhevnikova L.G., Petropavlovskii G.A. // *Zh. Prikl. Khimii,* 1993. V. 66.Vyp.3. P. 649 (in Russian).

[19] Sharpatyi V.A., / Chemical and biological kinetics. New horizons. V. 1. Chemical kinetics. Chapter 6. Ed.by: E.B.Burlakova, S.D.Varfolomeev, G.E.Zaikov, A.E.Shilov. Moscow: *Chimiya.* 2005. P. 108.

[20] Pristupa A.I., Sharpatyi V.A. // *Doklady RAN,* 2002. V. 387. V. 387., № 5. P. 643 (in Russian).

[21] Korotchenko K.A., Pristupa A.I., Sharpatyi V.A. // *Khimiya vysokikh energii,* 2004. V. 38. № 2. P.107 (in Russian).

[22] Nadzhafova M.A., Sharpatyi V.A. // *Doklady Akad. Nauk Azerbaidzh.,* 1977. V. 33. № 8. P. 41 (in Russian).

[23] Kvach N.M.,Kuvaldina E.V., Sadova S.F., Sharpatyi V.A. // *Doklady RAN,* 1996. V. 349. № 1. P. 60 (in Russian).

[24] Sharpatyi V.A., Shapilov A.A., Pintelin S.N. // *Khimicheskaya Fizika,* 2001. V. 20. № 12. P.19 (in Russian).

[25] Nishimura Yo., Kim Hee-S., Ikota N., Arima Hi., Bom Hee-S.,Kim Yo.- Ho, Watanabe Yo.,Yukawa M., Ozawa T.// *J. Radiat.Res,* 2003.№ 1. P.53.

[26] Park Pyo-J.,Je J-Y.,Kim S.-K.//*Carbohydrat Polym,* 2004. V. 55. P.17.

[27] Korotchenko K.A., Sharpatyi V.A. // *Radiation biologiya. Radioekologiya,* 2000. V. 40. № 2. P. 133 (in Russian).

[28] Burlakova E.B., Shishkina L.N. // The problems of natural and modifying radiosensitivity. / Ed. by M.M.Konstantionova and A.M. Kuzin. Moscow: *Nauka.* 1983. P. 29 (in Russian).

[29] Nechipurenko Yu.D., Volf A.M , Salyanov V. I., Evdokimov Yu.M. // *ZhETF,* 2004. V. 125. *Vyp.* 1. P. 103 (in Russian).

[30] Evdokimov Yu.M., Salyanov V. I., Krylov A.S.,Nechipurenko Yu.D., Volf A.M. // *Biofizika,* 2004. V. 49. *Vyp.* 5. P. 789 (in Russian).

[31] Mao Hq, Roy K, Troung-Le Vl, Janes Ka, Lin Ky, Wang Y, August Jt, Leong Kw // *J. Control Release,* 2001. V. 70. № 3. P. 399.

[32] Li Yh, Fan Mw., Bian Z., Chen Z., Zhang Q., Yang Hr. // *Chin. Med. J.* (Engl.), 2005. V. 118. № 11. P. 936.

[33] Prabaharan M., Mano Jf // *Drug Deliv.,* 2005. V. 12. № 1. P. 41.

[34] Eidus L.Kh. Phisiko-chemical principles of radiobiological processes and protection from irradiation. Moscow: *Atomizdat.* 1979. 216 p (in Russian).

[35] Kuzurman P.A., Sharpatyi V.A. // Radiation biologiya. *Radioekologiya,* 2006. V. 46. № 1. P. 22 (in Russian).

[36] Mosse I.B. Radiation and heredity: genetic aspects of radioprotection. Minsk: Universitetskoe, 1990. 205 p (in Russian).

[37] Mori T., Murakami M., Okumura M., Kadosawa T., Uede T., Fujinaga T. // *J. Vet. Med. Sci.,* 2005. V. 67. № 1. P. 51-56.

[38] Varfolomeev S.D., Gurevich K.G. Biokinetika. Moscow: *Grand,* 1999. 720 p (in Russian.

[39] Fang N., Chan V., Mao Hq., Leong Kw. // *Biomacromolecules,* 2001. V. 2. № 4. P. 1161.

[40] Paramonov D.V., Trofimov V.I., Knyazev A.A. // *Khimiya vysokikh energii,* 2004. V. 38. № 2. P.113 (in Russian).

[41] Han Y., Zhao L., Yu Z., Feng J., Yu Q. // *Int. Immunopharmocol.,* 2005. V. 5. № 10. P. 1533.

[42] Feng J., Zhao L., Yu Q. // *Biochem. Biophys. Res. Commun.,* 2004. V. 317. № 2. P. 414.

INDEX

A

B

C

D

Q

R

S